Precast Concrete

Also available from Taylor & Francis

Precast Concrete
Materials, manufacture, properties and usage

Second edition

M. Levitt

CRC Press
Taylor & Francis Group
Boca Raton London New York

CRC Press is an imprint of the
Taylor & Francis Group, an **informa** business
A TAYLOR & FRANCIS BOOK

CRC Press
Taylor & Francis Group
6000 Broken Sound Parkway NW, Suite 300
Boca Raton, FL 33487-2742

First issued in paperback 2019

ISBN-13: 978-0-415-26846-2 (hbk)
ISBN-13: 978-0-367-86403-3 (pbk)

Typeset in Sabon by
RefineCatch Limited, Bungay, Suffolk

British Library Cataloguing in Publication Data
A catalogue record for this book is available from the British Library

Library of Congress Cataloging in Publication Data
Levitt, M.
   Precast concrete : materials, manufacture, properties and usage /
M. Levitt. – 2nd ed.
      p. cm.
   1st ed. published : London ; Englewood, N.J. : Applied Science
Publishers Ltd., c1982.
   Includes bibliographical references and index.
   1. Precast concrete.   I. Title.
TA439.L443 2007
620.1'37 – dc22   2007007528

Visit the Taylor & Francis Web site at
http://www.taylorandfrancis.com

and the CRC Press Web site at
http://www.crcpress.com

To Evelyn, with love and thanks

Contents

Preface to the second edition

Over twenty years have passed since the publication of the first edition during which time there have been many changes in the UK precast concrete industry, of which the following are a few examples:

1   cessation of centrifugally spun processes for the production of pipes and lighting columns;
2   the implementation of European Harmonised Standards in the form of BS.ENs and BS.EN.ISOs;
3   the implementation of Eurocodes using mandatory instead of the recommendary clauses associated with the superseded Codes;
4   the likely implementation of the CE Mark on all products and systems placed on the market which, although common in most other Member States, had not become mandatory at the time of this book's submission for publication;
5   so many household names in precast ceasing to trade, resulting in changed emphases on types of products on offer;
6   large takeovers in both the precast and the cement industries resulting in larger dominances than used to obtain.

With all this in mind, the author decided that rather than update each of the first edition chapters a different approach was necessary. Targeting the main aim of readability, all but two of the old chapters have been re-ordered under new main and sub-headings but what have been deemed still applicable sections from the first book have been retained and updated as required. Although this created additional text, it saves the reader having to refer to the first edition unnecessarily. A new chapter dealing with finishing, jointing and repair techniques has been added as these subjects have been taking on an increasing importance of late.

It is hoped that this approach will be helpful to the reader and that the information given and the recommendations made will be appropriate for the current years as well as several years to come. In view of the author's advancing years, it is unlikely that he will be there to produce a third edition but it is hoped that someone with the same deep attachment to the precast concrete industry as the author will pick up these reins and ensure that an up-to-date edition will always be available.

Acknowledgements

The author expresses his thanks to John G. Richardson for his interest and helpful advice and the following who supplied data and photographs:

Mexboro Concrete Ltd
Roger Bullivant Ltd
Trent Concrete Ltd
The British Precast Concrete Federation
The United Kingdom Cast Stone Association

List of abbreviations

The following is a list of abbreviations and acronyms that will either be found in the text or are likely to be encountered by the reader in the references and/or in the many publications relevant to the concrete industry.

A/C	aggregate/cement
AA	accelerating admixture
AAC	autoclaved aerated concrete
AAR	alkali-aggregate reaction
AEA	air-entraining admixture
ALR	alkali-limestone reaction
ASR	alkali-silicate reaction
BS	British Standard
BS.EN	British Standard
BS.EN.ISO	British Standard
C	coarse
CAC	Aluminous cement or Calcium aluminate cement
CAC	ciment fondu
CEM I	Portland cement (95–100 per cent clinker)
CEM II	Portland-composite cement (65–94 per cent clinker)
CEM III	Portland blastfurnace cement (5–64 per cent clinker)
CP	Code of Practice
CS	compressive strength
dBA	sound reduction
E	Young's Modulus
EPS	expanded polystyrene
ETA	European Technical Approval
F/T	freeze/thaw
F	fire
FA	foaming admixture
FS	flexural strength
GGBS	granulated ground blastfurnace slag
GRE	glass-reinforced epoxide
GRP	glass-reinforced polyester

HAC	high alumina cement
HEN	harmonised Standard
HRWRA	superplasticising/high range water-reducing admixture
HWA	high density aggregate (e.g. barytes, ironstone)
ISAT	initial surface absorption test
LWA	low density aggregate (e.g. sintered PFA)
M	medium
MM	moisture movement
MPa	mega-Pascals
MS	microsilica
N	normal rate gain of strength
NWA	normal weight aggregate (e.g. flint, limestone, granite)
PFA	pulverised fuel ash
PVA	polyvinyl acetate
PVC	polyvinyl chloride
R	rapid strength gain
RA	retarding admixture
RHPC	rapid-hardening Portland cement
S	slag
SC	self-compacting
SG	specific gravity
T	temperature
TCMB	temperature matched curing bath
TS	tensile strength
U	thermal transmittance
W/C(F)	water/cement (cementitious) ratio, free
W/C(T)	water/cement (cementitious) ratio, total
WPA	waterproofing admixture
WRA	plasticising/water-reducing admixture
WRA	water-repellent admixture

Introduction

The precast concrete industry continues as a mixture of labour- and machine-intensive processes and there is no reason to think that there will be a future weighting to more of one than the other. The nature of the industry is such that there are two classes of products, repetitive and bespoke with products such as paving flags, kerbs, pipes, extruded planks/beams and roofing tiles falling into the first machine-intensive category and virtually all the others into the second group.

How aspects such as energy-saving, market demand, costs and sustainability (a buzzword that will rarely be seen again in this book as it has yet to be defined) will affect the industry is anyone's guess. It would probably be a good thing for both the in situ and precast concrete industries to be more proactive in self-promotion. This, as exemplified by the disproportionate effects of high alumina cement, chloride corrosion and alkali-aggregate reaction, seems to have shown a reaction that was hardly reactive, let alone proactive. Perhaps, compared to the old British Steel days, there might have been some relationship with public rather than private industry interest.

Irrespective of the production involved, all precasters are subject to European Standards, Directives and Regulations, and their customers likewise. These are reflected in the high costs of disposal of waste materials, shortages of natural aggregates, pressure to use recycled concrete and new types of synthetic materials as aggregate and recycled mixing water and the use of cements containing blends of other materials. This list is not exhaustive and the industry needs to be prepared to deal with many other alternatives. The only thing that needs to be borne in mind is that irrespective of how well concrete can tolerate the use of other sources of waste or co-product materials, each application has to be fully performance-assessed before acceptability.

Chapters 1–6 re-order the text from the first edition in that the materials aspects are now dealt with in Chapters 1–3, entitled ingredients, reinforcement, prestressing, hardware and moulds. Pulverised fuel ash (PFA) and pigments are now covered in the Chapter 1 as are all admixtures and additives.

The author, having made several visits to the Middle East, has used the experience gained there to deal with precasting in hot and cold climates (Chapter 8). This should be of interest to readers who either currently operate or intend to do so in these climes. Experience gained from Scandinavian and Finnish visits, coupled with the work undertaken on the RILEM freeze/thaw committee, has been used to give advice on the cold climate section of Chapter 8. The chapter follows the one on accelerated curing as there are many facets involved that relate to work in hot climates.

Chapters 9 and 10 cover properties and performance, Standards, testing and quality, and aesthetics. Even though aesthetics is a property/performance matter, it was deemed an important enough property to merit separate emphasis.

Chapter 11 covers a new group of subjects and discusses finishing, jointing and repair techniques. Finishing products is of long-standing application whereas in both repair and jointing there is quite a future for designers and precasters who can appreciate the viable combinations of concrete with plastics.

1 Ingredients

The first edition[1] of this book was a little off balance in comparing the amount of text on additives, admixtures and pigments with the main ingredients of cements and aggregates. The reason for this was that in 1982 there were a large number of publications on cements and aggregates which were referred to by a large readership as being applicable to all concrete production, including precast. However, this assumption cannot be made as the precast industry generally has more stringent demands on raw material requirements compared to in situ concrete. In view of the changes that have taken place in the precast concrete industry over more than two decades, coupled with stricter regimentation in the use of Standards and system approaches, what is considered to be a better balance has been adopted in this edition.

The Notes and references section is deliberately brief so that the reader can pick up the main points in the current text. This edition is in stand-alone form and the reader should not require to refer to the first edition.

In situ concrete producers, users and specifiers now are subject to the European Standard[2] and Code.[3] These documents do not apply to the vast majority of precast products as they are generally covered by specific Eurostandards. The Standard for concrete is supported by a complementary two-part British Standard[4] giving national guidance. The other matter to be borne in mind with virtually all Eurocodes is that they are written with a mixture of 'shoulds' and 'shalls' compared to the current or superseded recommended 'shoulds' of BS Codes.

1.1 Cements

To list all the types of cements covered in the various parts of the Eurostandards for cements[5] would take an inordinate amount of space. As only a few of these Standards are likely to be used in the precast concrete industry, only those cements will receive primary attention.

There are several types of cement covered in these Standards even including masonry cement. The Portland cements listed contain OPC varying from 100 per cent down to 5 per cent clinker content with CAC aluminous cement being the only one having 100 per cent clinker. Although CAC (originally known as high alumina cement, HAC, Ciment Fondu, Lightning Cement) received an unjustified[6] bad name in the 1960s, it is still considered to have useful potential in the precast concrete industry.

Cement designations are complex and many sub-classifications are used. A rapid-hardening Ordinary Portland Cement, (RHPC) could be designated by:

CEM I/A-mac.52.5N

where:
'CEM I' is the main (95–100 per cent clinker) cement type
'A' confirms a high clinker content and could be omitted
'mac' is an inorganic 'minor additional constituent' at 0–5 per cent
'52.5' is the 28-day characteristic cube strength which is specified to be
    52.5MPa minimum. If the '52.5' had been '42.5', the 28-day characteristic cube strength would have to lie between 42.5MPa and 62.5MPa
'N' is a normal rate gain of strength. If the 'N' had been 'R', standing for
    rapid strength gain, there would be no limits for the characteristic strength.

The characteristic strength is the figure below which not more than 5 per cent of the cube results would be expected to fall. In the Standard, the concrete strength covers both cube and cylinder testing, with the cylinder strength given first. Thus, a characteristic 25MPa cube strength would be specified as 'C20/25' as the cylinder's 2/1 slenderness ratio would result in a lower crushing load for the identical concrete in the 1/1 slenderness ratio cube.

White cement, commonly used for cast stone and architectural concrete manufacture, could be categorised as 'CEM I 42.5N' but would need the word 'White' somewhere in the description. The description could be made more complex in the case of a white Portland cement containing, say, 36 per cent GGBS where 'White CEM III/A-S 32.5N' could apply. The sub-class 'S' means that slag is the second constituent and the '32.5' limiting the characteristic strength to lie between 32.5 and 52.5 MPa at 28 days. The purchaser does not need to learn all the cement classifications by heart but should have a basic knowledge of those classes relevant to the work. This is necessary so that the most appropriate cement for the application be ordered and that the product, the delivery note and the corresponding cement test certificate all tally.

Setting and hardening times are terms that are applicable to cements and not concretes. The setting characteristics of the product are not the only properties relevant to the ability for early demoulding and handling. An equally important property for many of the machine-intensive processes is 'green strength', a property of the ability of retention of the moulded shape at a few seconds to a few minutes old. Probably only the fineness of the cement relates to green strength, most of this being a function of aggregate grading, cement fineness, water/cement ratio, free, W/C(F) and compaction efficacy.

1.1.1 Cement problems

The list of problems that follows the general preamble on cements refers to problems either encountered by the author or possible future ones. Both the in situ and precast concrete industries now have access to a far larger range of cements than used to be available. This, for most of the Portland cements, is largely due to the use of what may be termed inert extenders that are added to the clinker prior to grinding.

Generally speaking, the cements available fall into two fairly distinct groups:

1   Diluted clinkers containing a small percentage of materials, such as limestone, resulting in properties not significantly different from the undiluted version. With carbon emissions from industry needing to be minimised to help inhibit the impending global warming and 1T of cement manufacture said to produce 1T of carbon (carbon dioxide), users should condone clinker dilution. On the other hand, a degree of wariness should be exercised as other extenders may be used without full assessment (cement tests and applicability to the precast product in question).
2   Cements containing significant quantities of additives such as PFA, granulated ground blastfurnace slag (GGBS) and microsilica (MS) which are used either for the specific properties of the additive and/or for properties that arise from its reaction with the cement component.

The following is a list of problems that can arise:

1   The finer the cement, the faster the setting and hardening times and this makes the higher characteristic strength cements more attractive to the concrete industry, especially the precast side. The main downside to such selection is an enhanced rate of exotherm leading to too rapid a

curing rate, excess thermal gradient and cracking. In addition, some machine-intensive processes do not operate well with very fine cements. One process very sensitive to the fineness was spinning, not thought to be practised in the UK at present. The process was used for pipes and lighting columns and manufacturers used to order a special cement categorised as 'coarse ground' even though its specific surface complied with the then limit for Ordinary Portland.

2   Disputes often arise because of the cement and, if the precaster has not kept a representative sample, there is no proof as to where the blame lies one way or the other. Every delivery of cement should be appropriately sampled and stored in airtight containers labelled with all the relevant information relating to the delivery. The label should also identify what part(s) of the production that cement delivery relates to.

3   Reliance is often placed upon a sulfate-resisting cement being enough so that attention is not paid to the main properties that affect sulfate-resistance such as mix design, compaction and curing. Blaming the cement for poor performance of the concrete can be misplaced.

4   While the word 'misplaced' is fresh in the mind, the reader is referred to the text *Concrete Materials*,[7] where in, the 1960s and 1970s the misplaced blame culture against aluminous cement was taken to task. It was estimated there that the three main precasters involved in production of pretensioned prestressed units in the years 1946–1975 manufactured about 15 million individual units and these went into about 60,000 constructions. A housing estate was defined as a single construction. About 1000 of the constructions (not estates) were examined by consulting engineers and four of these were found wanting. This 4 out of the total of 60,000 was small and most of the troubles were based on insufficient column bearings for beams and conversion leading to increased permeability and access of aggressive chemicals (e.g. chloride stabiliser from woodwool slabs) leaching through to the prestressing wires.

5   If the cement is not stored properly, this can lead to caking if moisture has access or the cement is stored too long. This is more of a problem with sack storage rather than with silos as, with silos, the newest cement delivery has to wait for the lower and older cement to be used first. Bag storage needs to be rotated so that the oldest storage is used first, avoiding the problem of dumping the latest delivery on top of the older stock.

1.2 Aggregates

It is difficult to separate economic and political factors from technical and scientific considerations when there are so many pressures on both the aggregate supply situation (namely, sourcing and aggregate tax) as well as the disposal of waste materials (namely, the landfill tax). In South-east England flint gravel land supplies have been getting scarcer and scarcer and there has been a growing use of marine-dredged flint gravels and sands.

Concerning waste products, the current landfill tax, coupled with the costs of treatment and/or disposal of effluent makes it more and more attractive to recycle whenever possible. These and several other constraints relate to the discussions that follow.

It is difficult to divide aggregates into strict categories and the groupings adopted here are an attempt to keep this as simple as possible.

1.2.1 Normal weight aggregates (NWA) (typical SG range 2.0–2.6)

These can be grouped into natural and synthetic sources of material.
Natural:

(a) flint gravels and sands from land sources;
(b) marine-dredged flint gravels and sands;
(c) limestone crushed rock coarse and fines ('sands');
(d) sandstone crushed rock coarse and fines ('sands');
(e) intermediate sedimentary (c), (d) and dolomitic rocks;
(f) volcanic origin crushed rock and 'sands' (granite, basalt);
(g) limestone gravels and 'sands'.

Note: the word 'gravel' is often used to describe flints but it is strictly a geological term to describe all natural rounded angular aggregates and therefore applies to both flints and limestones.
Synthetic:

(h) calcined flint (used for architectural concrete);
(i) crushed concrete (this is discussed later);
(j) crushed clay or calcium silicate bricks(also grouped with crushed concrete).

Following a full assessment of aggregates (and combinations thereof) available, taking into account availability and transport costs, the precaster should be able to categorise orders for all his production requirements. As with cement supplies, the precaster should sample each approved ingredient appropriately and store in fully documented labelled airtight containers. There is no cause for complaint of change in material unless one had an original sample for comparison.

A few of the typical problems with aggregates that have been encountered are listed below:

1   Flint gravels from land sources can have a frost-prone cortex coating (a skin of material typically 0.5–2.0mm thick) which, if a concrete is subject to freeze/thaw (F/T) damage, can result in 'pop-out craters' over the larger pieces of aggregate near the surface.[8] The aggregate generally

has a higher thermal conductivity than the surrounding mortar and this will tend to cause a lower temperature in that area with resultant moisture concentration and a tendency to approach or exceed the critical water saturation level in the cortex. This damage is not to be confused with alkali-aggregate reaction (AAR) which, although a rare occurrence, shows up as massive cracking and expansion.

2   Marine-dredged materials need to be washed to remove most of the salt, sulfate and organic material, a certain amount of salt and shell being tolerable. It is recommended that if the salt content is to be measured, it is done on the washed unsieved as-received material as the sieving process results in salt sticking to the sieve meshes, giving a false and lower than true result. This could result in a non- compliance result if a sieved aggregate salt content figure was a marginal compliance whereas an unsieved figure would be a failure.

3   Crushed limestones have had a very good track record in the industry for nearly a century. Shape factor can be a drawback with any rock aggregate and this defect can possibly be traced to the supplier crushing through too few stages. Too much dust (often referred to as 'flour') can be a problem with 'sands'. In vibrated concrete, a higher water demand would be needed for a given workability but the limestone can cause a 'vacuum' suction effect enabling early demoulding. This suction results in a loss of workability and a stiffening of the mix, giving an acceptable demoulding green strength. In moist mix design cast stone, neither too much nor too little dust is needed in order to obtain the ex-mould natural stone appearance and good void filling coupled with a good green strength.

4   Coarse aggregates of crushed 'granite' and basaltic-type rocks can have the same shape problems as discussed in 3. Some sources are known as 'flakey' due to the way they are crushed. 'Sands' from these sources can be very dusty unless washed or spun to remove excess unwanted material but some machine-intensive processes need some very fine aggregate. For vibrated wet-cast products the shape drawback effect on workability can be counteracted either by the use of admixtures and/or the incorporation of some natural rounded sand.

5   Limestone gravels and 'sands' are often used in architectural products due to their attractive and 'warm' appearance. The presence of pyrites can cause unsightly staining and no known aggregate preparation method will remove these particles. Oxidation and staining tend to occur within the first few weeks and it is adviseable to promote this reaction before delivery by storing outdoors thus promoting oxydation of the iron sulfide. The offending pieces can then be cut out and the areas made good. Pyrites can easily be identified by placing the suspect aggregate in dilute hydrochloric acid and detecting the hydrogen sulfide bad egg smell. In addition to taking the usual precautions when handling acid, it needs to be remembered that the limestone

will react violently with the acid, emitting large quantities of carbon dioxide gas.

6   AAR is commonly associated with the presence of chert, opaline and similar reactive components in flint gravels. The reaction is between the caustic soda and caustic potash from the cement and aggregate components resulting in severe expansion and cracking, with cracks having surface widths of up to several millimetres. Some limestones have occluded clay within their depth which reacts with the cement alkali in a similar expansive reaction known as dedolomitisation. Therefore AAR has two sub-groups covering reactions with silicate and with carbonate limestone rocks and these should be given the initials ASR and ALR respectively. Fortunately, in the UK, ASR is quite rare and ALR is virtually unknown.

As the number of complaints compared to the total number of contracts is very small, one cannot help drawing a comparison with the high alumina cement scare of the 1970s. This is no excuse to dismiss any suspect hazard and, if tests indicate that such is the case, then the use of low alkali cement, or specific admixture and/or additive can be considered. Some tests for AAR can indicate the presence of reactive material but a long-term full-scale mortar bar expansion test is probably the only way to differentiate between reactive and dangerously reactive material presence.

The book by Smith and Collis[9] is recommended for an in-depth study of this topic.

1.2.2 Low density aggregate (LWA) (typical SG range 1.2–2.0)

At the time of the publication of the first edition there were about six sources of LWA with all the synthetic ones being of UK manufacture. From the author's current knowledge today only one LWA is in use and that is sintered PFA imported from a mainland European source of manufacture. The reason for this decline is unknown. It could be due to the manufacturing heat requirement, bearing in mind that, unlike cement manufacture, LWA constitutes a large proportional volume of the concrete mass. Another reason could be that there is a dearth of designers used to specifying and using LWA for structural purposes. One could then ask what is being used for lightweight concrete block production and the answer to this would appear to be autoclaved aerated blocks. This, in turn, would mean that typical natural aggregate concrete blocks with densities in the range 600–2000 kg/m^3 are not available.

An old and common source of LWA used to be clinker and this might well return in future. (Note: the term 'clinker blocks' is still in common use to describe concrete blocks.) The UK still has large coal reserves and it is possible that the industry will be reinstated for the production of electricity, gas and fuel oil. The by-product is clinker with no energy of embodiment as is the case with the synthetic materials. It should also be borne in mind that coal dust used to be used to fire cement kilns and this fuel could become competitive in future.

As with NWA, there are two groups with only one in the natural group as far as is known. Those listed in the synthetic group, apart from sintered PFA, are not currently in production.

Natural:

(a) pumice – a porous and permeable volcanic rock;
(b) perlite – similar but a very friable porous obsidian volcanic rock.

Synthetic:

(c) expanded clay – a fired porous particle, rounded in shape;
(d) expanded plastics – typically pellets of expanded polystyrene;
(e) expanded shale – a high angularity particle with low porosity;
(f) expanded slate – as expanded shale;
(g) foamed slag – a by-product from steel production;
(h) sintered PFA – pellets of PFA fired on a sinterstrand. The particles have a high water absorption.

Under the problems listed below have been grouped aspects of general information.

1   The two sources of pumice that have been used are the darker material from Finland and the lighter one from the Italian islands. The Finnish source is a mountain in a very windy area and the finer fraction tends to become deposited further down the slope. This enables the supplier to sample gradings according to the distance from the crater. The Italian source material has to be sieved or used as an all-in aggregate. Both pumices are relatively weak in crushing strength and are therefore probably only of application as full or part aggregate in concrete block production.

2   The main problem with clinker aggregate used to be pop-out craters in the finished product due to the slaking of quicklime. The quicklime was produced in the firing process where pieces of chalk became converted. The CaO quicklime presence is not obvious to the eye but is quick to react with moisture with accompanying large expansion. The effect can be largely counteracted by keeping the stockpile wet for 2–3 weeks prior to use relying upon wet weather or, in dry weather, by artificial means.

3   Sintered PFA, as well as pumice and expanded clay, has a high water demand, resulting in large margins between W/C(T) and W/C(F) and this needs to be catered for in designing the mix. However, in a similar vein to limestone mixes, this vacuum suction property can be used to

promote green strength enabling demoulding before any significant setting takes place. This requires strict control of the W/C(T) in the works allowing for the variabilities that will occur in batch-to-batch deliveries.

4 Concrete strengths should always be designed to the capability of the aggregate; it is of no avail aiming for a characteristic strength of 40MPa when the crushing strength of the aggregate is 50MPa. In addition to this, it needs to be borne in mind that cube strengths are unlikely to be Gaussian in shape and will tend to resemble a reverse Poisson shape due to the higher end rapid tail-off.

For further in-depth reading on this topic, refer to Clarke[10].

1.2.3 *High density aggregate (HWA) (typical SG range 3.5–6.0)*

These aggregates have an application in concrete for radiation shielding, airborne sound insulation of walls and partitions, counterweights for tower cranes, forklift trucks, spin driers and encapsulation of hazardous materials. Their application in the precast industry is not particularly significant but, if the current government thinking on the future of the nuclear industry for the generation of electricity is accepted, a resurgence in demand may be expected. Precast products for a nuclear industry are also likely to be specified to very tight tolerance for their close assembly needs.

The current choice of HWA is not large but, with pressure on the concrete industry to use other aggregates, it may be expected to increase. As with all developments in materials, a thorough assessment of suitability should be made in all cases. Such assessments, as with suspect AAR, should not be limited to short-term tests. Some harmful mechanisms, both chemical and physical in nature, may only be identified over a relatively long period.

There are two natural sources and one synthetic known to the industry currently:
Natural:

(a) barytes – a barium sulfate mineral at the lower end of the density range;
(b) ironstone – rock containing iron-bearing ore.

Synthetic

(c) iron, usually in the form of lumps.

The following is a list of problems encountered:

1 Barium sulfate, barytes, can also contain soluble sulfates which can retard setting and hardening times as well as affecting the sulfate resistance of the hardened concrete.
2 Obviously the same volume of HWA concrete will weigh considerably

more than NWA or LWA concrete and the ability of the mixer and its supporting structure to withstand the extra load should be assessed.

3   Lumps of iron or ironstone being struck by moving parts of the mixer then striking scraper blades and similar can cause damage, and not only can a HWA concrete process benefit by mixing at a slower speed than normal but mixer parts might need a more robust design.

4   Iron-containing aggregate will cause surface rusting if products are stored outdoors and should be kept under cover or be surface-protected in some way. The author has no knowledge of what might be suitable and suggests, without recommending, a phosphoric acid trestment to the clean and dry surfaces followed by a pigmented epoxide-based paint.

1.2.4 Crushed concrete

Under this heading one can include crushed clay and sandlime bricks as one cannot be sure of the aggregate being always specific, bearing in mind that site waste materials will be crushed en masse. The main problems with the aggregate preparation are crushing and screening and there are now mobile plants that can be taken to the demolition or other area if there is room for them. The main problems with the aggregate in concrete are:

1   an increased moisture movement characteristic of the hardened concrete;
2   a contribution of additional sulfate, and possibly chloride, to the mix, affecting the setting and hardened properties.

Initial trials reported by HMSO[11] reported a generally satisfactory assessment and the aggregate was used in the construction of a building at the Building Research Establishment at Watford. From information gathered it seemed that the aggregate was only used in the footings concrete and not above ground. This possibly questions the confidence placed in the findings as well as the logic of the application as one would like to observe the concrete visually to see its behaviour with time. If repairs or replacement were necessary, above-ground concrete would be more amenable to such work than footings.

1.2.5 Gradings

It is best to store coarse aggregates in their single sizes, thus facilitating selection for mix design. For all-in aggregates care needs to be exercised by watching the tail-end effect of the fines. The aim of any grading selection, apart from concrete block production, is to obtain minimum voidage structure by filling up as much of the void structure as possible. This is normally obtained by having a mixture of the largest size, an intermediate size and sand or fines. Tables 1.1 and 1.2 show typical coarse and fines gradings.

Table 1.1 Nominal single size coarse aggregates

Size	Passing sieve					
	63 mm	*37.5 mm*	*20 mm*	*14 mm*	*10 mm*	*5 mm*
63	85–100	0–30	0–5			
40	100	85–100	0–25	0–5		
20	100	100	85–100		0–25	0–5
10	100	100	100	100	85–100	0–25

Table 1.2 Suitable fine aggregate gradings (natural sands)

Type	Passing sieve					
	5 mm	*2.36 mm*	*1.18 mm*	*600 μm*	*300 μm*	*150 μm*
1	90–100	60–95	30–70	15–34	5–20	0–10
2	90–100	75–100	55–90	35–59	8–30	0–10
3	90–100	85–100	75–100	60–79	12–40	0–10
4	95–100	95–100	90–100	80–100	15–50	0–15

However, depending upon the precast process in question, these are not 'tablets of stone' and the precaster needs to lay down to the supplier what suits his requirements and ensure that they are regularly consigned.

Sands and crushed rock fines (often called 'sands') are generally specified as coarse (C), medium (M) or fine (F) with the acceptable grading range much wider than the superseded 'types 1–4' that used to be specified. With a high degree of latitude '1' is 'C', '2' and '3' are 'M' and '4' is 'F', all with a degree of overlap.

It should also be remembered than silt, as defined by the passing 75 micron fraction, is not the same for natural sands as for crushed rock material. In natural sands, it is often reactive clay whereas, for crushed rock, it is geologically identical to the larger fractions.

Depending upon the precast process, it is dangerous to make a generalisation that silt should be at a minimum. As an example of a specific requirement, for moist mix design cast stone production, an idealised grading would be equal proportions in each sieve range starting with the 2.36mm size. Rocks, if crushed in proper form, tend to produce such gradings but would probably need most of the passing 150 micron sieve removed for structural concrete application.

1.2.6 Shape

Some crushed rock sources are often denigrated for being too flakey when a closer examination of the processing might well reveal that only two stages of gyratory crushing is used where three or four stages would be better for the production of most concreting aggregates. From the quarry the initial stone sizes can be as large as 150mm in size and the crushing process has to address the hardness of that stone. The word 'most' was used above as the facing mix in wet/dry pressed floor tiles requires a high flakiness index aggregate.

Table 1.3 compares a marine-source sand with Middle East desert sands. The two lessons to be learnt here is that marine sands, being washed, are generally fines-deficient and this might not suit some precast processes. Second, precasters venturing into desert climes may well need to be warned to take special care in aggregate dustiness, especially when water is expensive and in short supply.

1.3 Water

The cheapest ingredient is often taken for granted and its properties ignored, concentration only being placed upon the quantity used and its macro effects. It is often said that, putting the test requirements aside, water fit enough for drinking is suitable for any concrete work. This is not always true and a concrete block manufacturer is known to use filtered sea water for mixing. Apart from excess lime bloom, a tendency to a hydrophilic nature and promoting rust for most metals getting into contact with the concrete, the blocks were used satisfactorily.

Recycled filtered mix water from wet-pressed processes has been used with no significant problems for many years and other processes, such as washing exposed aggregate products, can also generate a useful source of water. As with the earlier discussed aggregate possible shortages, pressure to use water in all industries more efficiently is current.

The bottom line of all mix water suitability queries is to make the product and assess its properties compared to the same product made from mains water. There are likely to be differences but it is up to the manufacturer to determine whether or not these differences are acceptable.

Table 1.3 Other natural sand gradings (not necessarily typical)

Source	Passing sieve					
	5 mm	2.36 mm	1.18 mm	600 μm	300 μm	150 μm
UK – marine	100	100	100	92	8	2
Saudi – dune	100	100	100	89	45	19
Jordan – wadi	97	95	92	86	60	28

1.4 Admixtures

Before going into this subject in depth and at the risk of being accused of being anecdotal, perhaps a true story about an incident that took place at the then Cement & Concrete Association (C&CA) about 1958 will illustrate that there is nothing new in the use of admixtures for cementitious mixes. They were in use over 2000 years ago.

In a road development on what was believed to be the old Roman road at Fenny Stratford (in the Midlands), some paving was exposed set in an old mortar. The site became an archaeological 'dig' and was identified as of Roman origin. Knowing the expertise of the Romans in producing cementitious compounds, a sample was sent to the C&CA for chemical analysis.

The routine for analysis involves first of all reacting a sample of the ground material with hydrochloric acid. This left a fatty-looking film on the surface of the acid which seemed to be of organic origin. A sub-sample of the mortar was sent to the Paint Research Association for identification of the organic part but they failed to identify this but opined that it could be of a bio origin. Another sample was sent to the Medical Research Council and they identified it as a sulfonated haemoglobin.

In effect, it appeared that the Romans were treating blood (assumed from animals but could have been from another source) with sulfuric acid to produce a sulfonate-type workability aid for mortar mixing. The process was rumoured to have taken place in Russia in the 1960s using blood from slaughterhouses as the organic radical but with the same aim of producing an admixture.

Admixture manufacture is now a factory synthetic process with probably more than 70 per cent of all UK concrete containing one form or another of an admixture. There has been a vast swing in usage over the past 30 or so years when, due to a general resistance to use, possibly 70 per cent or more of the concrete produced had no admixture. The commonly used workability aids have the main advantage of producing an identical workability for a lower water content. The second advantage is that less water is used which is the aim in a water-shortage era.

What this means is that the old adage that a well-designed concrete mix does not need an admixture is not true and was not true at the time it was stated. There are many desirable properties in both the fresh and the hardened states that can only be obtained by the use of admixtures that one wonders why they were not used widely in the early days of concrete. The author appreciated this need in his cast stone development work in the 1950s and 1960s and was the instigator of 'Waterproof Snowcrete', a commercially available cement bestowing water repellent properties on the product.

The main trap awaiting the unwary is similar to the earlier comment made regarding sulfate-resisting cements. Because an admixture gives desirable extra properties is no excuse for not designing and making the concrete to the best of one's ability. Admixtures do not make mediocre concrete good nor make bad concrete mediocre. Another trap is that they are all generally used at low concentrations, typically 0.01–0.5 per cent m/m cement which, in turn, means using precision and regularly calibrated or validated dispensing equipment. For liquid admixtures there is a large range of dispensers one can buy, whereas powder dispensers are not so common. Some powder admixture manufacturers of 'waterproofers' appreciated the dispensing problems over 50 years ago and still supply the active product diluted with an inert filler so that the concentration requirement of, say, 0.1 per cent m/m cement is achieved by manually adding 1 per cent m/m cement of the proprietary product (e.g. 1 bag per 50kg bag of cement).

For further information, the Concrete Society has several publications and data sheets led by a general publication.[12] For a full appreciation of the subject, refer to Ramachandran[13] as a full treatise; this text not only discusses admixtures but also includes additives and pigments. However, concentrating on the precast industry, there are two main groups, each with several sub-groups:

Group A Workability:

al   water reducing;
a2   workability increasing;
a3   a1 + a2 combined;
a4   self-compacting (full flow/collapse slump).

Group B 'Waterproofers' (in quotes because some cannot withstand high water pressures):

b1   water repellents;
b2   waterproofers.

It will be noted that AEA have been omitted because they are not known to the author to be in use in precast products. Since only wet-cast vibrated products could be candidates for frost resistance, these admixtures' non-use is probably because precast products generally obtain their frost resistance through a higher cement content than in situ concrete.

Frost resistance obtained through air entrainment is not a function of the amount of air entrained but of the bubble size distribution and bubble spacing. These factors are sensitive to the fines grading and cement content and, for a cement content above $350kg/m^3$ it becomes difficult to entrain enough of the right sort of air. Cement contents for wet-cast vibrated products would typically be $400kg/m^3$ or higher.

Before going into these in detail it is useful to say something about 'accelerators', a term which is often used to describe Group A products but formerly was used for chemically acting accelerators. In this first description for Group A the term only reflects on water reduction, resulting in an increased speed of hardening with no significant accompanying increased speed of setting. The second description is chemically the correct one. Calcium chloride used to be a commonly used all-rounder as it accelerated setting and hardening times for all Portland based cements. Its popularity was dashed by a degree of over-exuberance mainly in the precast industry, resulting in corrosion of reinforcement with accompany cracking and spalling. However, the products produced by the less exuberant precasters who exercised strict control are still performing well. In addition, the level of confidence in calcium chloride was so high in the 1950s and 1960s that it was possible to purchase '417 Cement' which had 1 per cent m/m anhydrous calcium chloride mixed into the cement.

The other chemical accelerators such as calcium formates and phosphites do not seem to perform well for the whole range of Portland cements and appear to have sensitivity to the tricalciumaluminate content of the cement as shown in Table 1.4.

Although a great deal of work on calcium chloride admixture was undertaken and reported at length in the first edition,[14] in view of its demise, these data are not discussed in this edition. The one exception was work carried out on low concentrations of calcium chloride. The reason for that study was to see if very low concentrations could cause anomalous strengths, a well-known but rare occurrence in cube testing. Tables 1.5 and 1.6 illustrate

Table 1.4 Comparative cube strength typical comparisons with calcium formate admixture at 1% m/m cement

C3A in cement (%)	Age			
	24 h	*3 day*	*7 day*	*28 day*
14	1.0	1.1	1.0	1.0
9	1.4	1.4	1.3	1.3
3	1.9	1.8	1.6	1.5

Table 1.5 Ratio to control strength for 24 hour old 4/1 mortar cylinders

Anhydrous CaCl₂/cement (w/w)	Total W/C						
	0.4	*0.5*	*0.6*	*0.7*	*0.8*	*0.9*	*1.0*
1.0	2.1	1.6	3.5	2.1	2.5	1.6	1.7
0.5	1.8	1.4	3.1	1.8	2.2	1.9	1.4
0.1	1.5	1.5	2.6	1.6	2.0	1.5	1.2
0.05	2.0	1.6	2.9	1.6	1.8	1.3	1.1
0.01	0.9	1.3	1.8	1.1	1.7	1.9	2.0
0.005	0.6	0.5	0.8	0.9	0.8	1.0	1.7
0.001	0.7	0.5	0.8	0.6	0.7	0.7	0.5
0.0005	0.8	0.5	0.8	0.7	0.7	0.9	0.5

Table 1.6 24 hour old and later age comparative strengths for 4/1, 0.5 W/C(T) mortars

Anhydrous CaCl₂/cement (w/w)	Age			
	24 h	*3 day*	*7 day*	*28 day*
1.0	1.6	1.5	1.3	1.1
0.5	1.4	1.4	1.3	1.0
0.1	1.5	1.4	1.2	1.0
0.05	1.6	1.4	1.3	1.1
0.01	1.3	1.2	1.2	0.9
0.005	0.5	0.5	0.6	0.7
0.001	0.5	0.6	0.6	0.6
0.0005	0.5	0.5	0.6	0.7

the behaviour for 28-day old mortars at different W/C(T) ratios and at different ages for mortars at a W/C(T) of 0.5.

It may be seen that below 0.01 per cent m/m cement anhydrous calcium chloride anomalies begin to appear. The question might arise as to who would want to add chloride at such a low concentration but this could easily emanate from chloride in the aggregate or other sources dissolving into the mixing water. The mechanism of chloride retardation is unknown and could be associated with a buffering reaction. Whatever the cause, it could be the reason for the 30–40 per cent decrease in the 28-day mortar strengths observed and for the rare outriders in laboratory cube testing in otherwise perfect cubes. Furthermore, it is not known if, for the particular cement used, there is not a more critical concentration of chloride giving more retardation, nor what the triggering concentrations are for different cement chemical compositions.

1.4.1 Workability

Chemicals based upon lignosulfonate, carboxylic acid, naphthosulfonate, formaldehyde and similar are used in concentrations ranging from 100ml to 1000ml per 50kg of cement. The range is doubled for the high range water-reducing (HRWR) and self-compacting (SCC) concretes. There is some evidence that the choice of the type of plasticiser added should relate to the grading of the fines with coarser sands preferring lignosulfonates and finer ones carboxylic. Trial mixes should always be assessed to find out which one suits a particular production. The lower end of this range applies to the lower end of workability improvement and the higher end to HRWR and SCC mixes. These admixtures are (or should be) dispensed at the same time as the mixing water and with the mixing water into the mixer. Users are warned not to add these admixtures into the dry mix ingredients as they can coat cement and cementitious particles, inhibiting the hydration and causing retardation.

The mechanism of working is that negative charges are placed upon the particle surfaces causing them to repel each other. With the naphthosulfonate and formaldehyde admixtures a much stronger dipole is formed with the water molecules giving the workability effect a much shorter life than with lignosulfonates and carboxylics.

As noted earlier, the three uses of workability aids are to improve workability at the control W/C(T), to maintain the control workability at a reduced W/C(T) or a combination thereof. There is a fourth advantage and that is an improved surface finish with either fewer and/or smaller surface air holes which is desirable for visual concrete finishes.

All applications need to be thoroughly assessed before being used in full production and, like aggregates and cements, fully recorded details on representative sealed samples should always be retained. It is advisable in assessment to test for air entrainment as there might be a risk of this happening in some combinations. If concrete is to be accelerated by using heat, the same rule applies in the assessment. It could be pointless assessing at room temperature and then curing at, say, 60°C. The necessary health and safety requirements also need to be observed bearing in mind that some operatives might be sensitive to the chemicals in use.

1.4.2 Superplasticisers and self-compacting admixtures

At the higher concentration mentioned earlier and more often using naphthosulfonate and formaldehyde-based admixtures, HRWR and SCC properties can be achieved. As already mentioned, a strong dipole moment is formed. This tends to override the electrostatic repulsion rather quickly and at about 20°C the effect can be lost at 1 hour after admixture addition. This is why readymix concrete trucks add the admixture on site just before offloading.

HRWR and SCC admixtures can be used in the a1 style where very high early strengths can be obtained. It is thus possible to use low workability (e.g. 50mm slump) concrete with W/C(T) down to about 0.35 by these means. However, depending upon the mix temperature, the mix would become unworkable within a short period. Therefore, HRWR and SCC admixtures should not be used in heated mixes and extra care needs to be exercised for ordinary mixes during hot weather.

SCC concrete in deep fills such as precast ducts, beams and similar needs to be treated as a high specific gravity (SG) liquid with the resulting hydraulic pressure capable of being withstood by the mould design. In addition, mould joints need to be very tight as these free-flowing mixes can easily lose grout at these areas.

The general promotional aspect of workability aids being defined as 'waterproofers' needs to be taken in context. They do not make the concrete waterproof but improve the permeability due to the better workability, enabling better compaction and an overall smaller pore size.

1.4.3 Waterproofers and water repellents

True waterproofers are products that stop the passage of water through the matrix, maintaining their performance under high hydrostatic prsssure. Since concrete virtually always has a pore structure and the solid material particles that surround the pore structure can absorb and adsorb water, a waterproofer has to do act in dual mode:

1   it has to block the pore structure effectively;
2   it has to coat all the surrounding solid material particles with a strongly effective water repellent layer.

Proprietary systems can achieve this dual target by a mixture of a swelling water-activated bituminous compound and a stearate-based powder that reacts with the free lime to form the water repellent metallic calcium stearate. A particular proprietary product has been extensively tested and used in many projects in Australia and the Far East. The author was associated with UK trials in basement work where the concrete neither cracked nor leaked and, following a long rainy period on site, the basements had to be pumped out before further work could continue. It is also worthy of mention that unpublished research by the author was conducted in the 1960s on a range of bituminous emulsions under the generic name of 'Piccopale Resins' which gave promising results for cast stone but was not extended to cover wet-cast vibrated concrete.

Water repellents are generally based upon stearates or metallic stearates and their mechanism is to line the pores and capillaries with a hydrophobic layer that repels water. The metallic soap is the repellent not the soap itself and stearic acid, usually in the form of a fine white powder, is only effective in wet-cast mixes where it can react with the free lime from the cementitious components to form the metallic soap.

In earth-moist mixes, it needs to be added as aluminium or calcium stearate because if it were added as stearic acid, the saponification reaction would occur over a very long time.

The main advantage of concrete or cast stone or many other products being made water repellent is that visual concrete keeps its appearance almost pristine for many years. Although there is a retardation in strength effect in wet-cast concrete, one advantage of earth-moist mixes is that the presence of the water repellent helps in moisture curing and promoting higher strengths. Table 1.7 illustrates this comparison for a concrete and a cast stone mix.

One problem with water repellent products is mortar jointing as the surfaces repel the mortar. This can be overcome by gauging the mortar mix with about 1 per cent cellulose solution instead of plain water.

As far as water repellent efficacy is concerned, placing a drop of water on the surface and observing a mercurial-type globulation is only an indication

Table 1.7 Control cube strength comparisons for calcium stearate admixture

Mix	W/C(T)	W/C(F)	24 h	7 day	28 day	Aggregates
20 mm/ 350 cement	0.55	0.47	0.8	0.7	0.7	Granite/ sand
5 mm/ 400 cement	0.45	0.32	1.3	2.0	2.4	Limestone/ sand

that it contains a water repellent, not how effective it is. The hydrostatic head of pressure of a water droplet represents little resistance to rain or rain-borne wind.

A test such as the ISAT (Initial Surface Absorption Test) should be undertaken where, at 200mm water pressure, satisfactory performance would be indicative of good performance in the worst of UK weathers.

More can be read about water repellents in cast stone in Chapter 5 on labour-intensive processes.

1.5 Additives

It is only 20 years since materials such as PFA, GGBS and MS became respectable and were given the generic name of additives. They were and are processed by-products or co-products from other industries which became respectable when it was realised that they had properties which made them very useful in concrete mixes, both for their chemical as well as their physical properties. The Romans and Greeks were using volcanic-source materials mixed with lime to produce their own concrete-like materials over two thousand years ago. It is surprising to think that a 100-year-old coal-fired electricity industry took so long to appreciate that a waste product could be so useful. Knowing how pozzolanic materials behaved, one would have thought that the slag waste product from steel production would have been examined for similar properties. Not finding pozzolanic properties in slag would have revealed the now well-known catalystic hydration behaviour of ground slag in the presence of lime.

The more up-to-date silicon production producing condensed silica fume was exploited in its early stages as it was found to be one of the most reactive (with lime) materials known. In an even more modern mode there have been attempts to get the concrete industry interested in other materials such as burnt shale, magnesium oxide and other products. As emphasised earlier, the industry needs to keep an open mind and only conduct full assessments when the supply situation has also been thoroughly investigated. Single source (geographically) materials should be regarded warily; users should generally have at least two alternatives for supply.

1.5.1 PFA

Powdered coal is fired to produce the heat required to generate electricity from steam generators. The powdered ash resulting ascends the chimney of the power station and is collected by electrostatic precipitators and stored in silos or bins. It can either be used direct being dispatched in tankers as cement is delivered or, after processing being made more reactive by air elutriation, collection of the most reactive particles.

The PFA particle is a nominally spherical hollow glass ball and its chemical make-up can vary quite widely depending upon the source of the coal, firing efficacy and other factors. Table 1.8 gives some values.

The ignition loss high numbers would only obtain from standby station. Modern power stations tend to produce PFAs with carbon contents of the order of 1 per cent m/m. Depending upon the carbon particle shape and size, carbon, being black, will result in a darker and darker ash the higher its content. Figure 1.1 illustrates this for a range of carbon contents. The colour varies from light to dark grey. The first four compounds in Table 1.8 are those that form the glass hollow spheres with the carbon sitting as discrete black particles in between.

The sulfate content can detrimentally affect the performance of products if at too high a level. Most products can tolerate about 5 per cent sulfate as SO_3 m/m cement and, as some cements can have up to 3 per cent m/m SO_3, no more than 2 per cent can be tolerated from the PFA. This means that if a PFA had 2 per cent m/m SO_3, it should not be used at more than equal weight with the cement.

The lime content can be in the form of free quicklime which, if present in a high enough quantity, can react with moisture and the silicate radicals in the ash to cause self-hardening. The magnesia MgO content would only contribute to expansion if it is in the form of periclase but this is not a common form of magnesia in PFA.

PFA has approximately half the bulk density of cement and is a dust hazard in the works and appropriate precautions should be taken. Its particle size ranges from 200–800 m^2/kg but, by selection of known suppliers,

Table 1.8 Ranges of chemical constituents in PFAs

Chemical	Percentage range
Silica (SiO_2)	28–51
Alumina (Al_2O_3)	12–34
Ferric oxide (Fe_2O_3)	4–26
Lime (CaO)	1–10
Magnesia (MgO)	0.5–2
Sulfate (SO_3)	0.2–3.6
Alkalis (R_2O)	1–5
Ignition loss – mainly carbon	1–32

Figure 1.1 PFA colour variations with carbon content.

a more restrictive range is possible as most precast processes prefer to work with ashes in the range 300–600 m²/kg.

For general reading on both PFA and GGBS, reference may be made to the Concrete Society's Report[15] with Swamy's[16] and Ramachandran's books[17] going into more detail. Having dealt with the generalities of PFA, the discussion will now turn to PFA's use in the precast concrete industry.

The long-term pozzolanic and improved sulfate resistance effects for in situ concrete are of little or no application for precast concrete products. PFA has one benefit for ordinary wet-cast products and that is the improved strength from pozzolanic reaction for hot concrete. The main fortes of PFA lie in the application to machine-intensive processes where one or more of the following benefits apply:

1 compensates for fines deficiency in natural sands;
2 rounded shape contributes to workability;
3 fineness contributes to green strength;
4 gives less wear and tear on plant and moulds;
5 inhibits lime bloom;
6 improves surface finish.

Wet-pressed paving and kerb

At the time (circa 1963) of the author's initial investigation into PFA in these products, several precasters were already using ash with mix designs established by works trials. What had been found in the works was that PFA loadings up to a certain level improved the strength (flexural) but that these improvements were too early for a pozzalanic effect. In addition, PFA loadings into concrete for the effect of pozzolanicity were typically at 20–40 per cent m/m cement whereas, for wet-pressed products the loadings were at 50–150 per cent m/m cement.

In order to investigate the mechanism in wet-pressed products, three ashes were obtained from a range of power stations that, at that time, exemplified those available. Table 1.9 shows the three main properties of these ashes F1, F2 and F3.

Each ash was sent to a separate wet-pressed works and kerbs were made with an ash/cement (A/C) of 5.9 of granite down to dust using OPC (Type I cement) with ash added at 0.25, 0.50, 1.00, 1.50 and 2.0 m/m cement and water added to suit the wet press process with the total amount recorded. Three kerbs were taken at 7 days old and three at 28 days old and flexurally tested and the average of each three results recorded.

Table 1.10 shows the initial water cement ratio and the ash/cement ratios as F/C and, using the carbon figures from Table 1.9, the carbon/cement (C) ratios.

Table 1.9 Properties of the three PFAs used in the works tests

Station source	Modern† F1	Old F2	Standby F3
Specific surface (m²/kg)	330	365	‡
Carbon content (C)	3.5	5.5	16.6
Sulfate content (SO₃)	0.8	1.2	1.8

† 'Modern' in 1963 when these ashes were sampled is no reflection on the later and improved boilers where a typical carbon content would be 1% or less.
‡ The standby ash could not be air-permeability tested as its high carbon content did not enable one to make a bed in the cell.

Table 1.10 Wet-pressed PFA mixes

A 5.9	C 1.0	Initial water 1.00	F1/C	Carbon/C	F2/C	Carbon/C	F3/C	Carbon/C
5.9	1.0	1.00	–	–	–	–	–	–
5.9	1.0	1.05	0.25	0.87	0.25	1.38	0.25	4.15
5.9	1.0	1.10	0.50	1.75	0.50	2.75	0.50	8.30
5.9	1.0	1.20	1.00	3.50	1.00	5.50	1.00	16.60
5.9	1.0	1.25	1.50	5.25	1.50	8.25	1.50	24.90
5.9	1.0	1.30	2.00	7.00	2.00	11.00	2.00	33.20

Table 1.11 lists the 7- and 28-day strengths with, in brackets, a strength correction for the total material cost. Taking the last entry in Table 1.10, the 5.9/1.0/1.3 (about 0.7 after pressing)/2.0 mix is about 20 per cent cheaper than the control mix with an estimated saving of 1 per cent for every 0.1 PFA/C increment. This gives an idea of the cost per unit strength.

Table 1.9 shows a wide range of properties and the sulfate levels can be generally dismissed as the cause of expansive disruptive effects as these would not be expected at ages up to 28 days.

A general plot of ash content against strength is not shown as the randomly distributed points do not show a best fit for any particular shape. However, when one plots carbon/cement ratios against 7- and 28-day strengths for all three factories and all three ashes the data begins to sit in a particular form. This is shown in Figure 1.2.

It may be seen that strengths increase at first then continously decrease and, at about 4 per cent m/m carbon/cement, the strength equals the control and at about 2 per cent m/m carbon/cement there is a maximum.

This means that provided the mix is acceptable to the process a PFA/C ratio of 2/1 of a PFA with a 1 per cent carbon content will give the best improved strength. However, depending upon the aggregate grading, the cement and PFA finenesses and other factors, a heavy loading of PFA might cause too long a build-up of pressure and hold times, making it uneconomic.

The improvement in strength would appear to be physical in nature and not chemical as each of the works not only reported the 7- and 28-day flexural strength improvements but also that the green strength, surface appearance and workabilities were all improved. There was some criticism of F3 in that the heightened darker appearance due to the high carbon content was not a good selling point. This leads one to the interim conclusion that the PFA produces an easier to compact mix and promotes a better filling of the void system. It would have been of little avail investigating this effect by density comparisons because the lower density of PFA compared to cement would likely have shown a decreasing density with improving strength. Samples of all the kerbs were submitted to the Initial

Table 1.11 Observed and cost-corrected flexural strengths of wet-pressed kerbs

F/C	F1		F2		F3	
	7 day	*28 day*	*7 day*	*28 day*	*7 day*	*28 day*
0	5.0	5.4	5.0	5.4	5.0	5.4
0.25	5.0(5.1)	5.7(5.8)	5.6(5.7)	6.0(6.2)	4.3(4.4)	4.8(4.9)
0.50	6.1(6.4)	5.2(5.5)	5.0(5.3)	5.2(5.5)	4.4(4.6)	3.8(4.0)
1.00	5.2(5.7)	5.5(6.1)	4.7(5.2)	4.7(5.2)	3.4(3.7)	4.1(4.5)
1.50	4.9(5.6)	4.7(5.4)	4.0(4.6)	4.0(4.6)	3.1(3.6)	3.7(4.3)
2.00	3.4(4.1)	4.1(4.9)	3.6(4.3)	3.6(4.3)	3.1(3.7)	2.8(3.4)

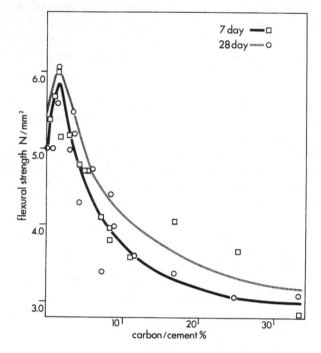

Figure 1.2 7- and 28-day wet-pressed kerb strengths vs. carbon/cement.

Surface Absorption test and the average results per trio of test samples are recorded in Table 1.12 in units of ml/m².s at the stated times from the start of each test.

A graph of the 10-minute results (not corrected for cost as Table 1.11) against carbon content is shown in Figure 1.3.

This indicates the same initial improvement followed by a worsening permeability as for the flexural strengths in Figure 1.2. The two trends are not as distinct as Figure 1.2 but it needs to be emphasised that the test is very

Table 1.12 ISATs of wet-pressed kerbs

F/C	F1			F2			F3		
	I 10 min	*I* 30 min	*I* 60 min	*I* 10 min	*I* 30 min	*I* 60 min	*I* 10 min	*I* 30 min	*I* 60 min
0	0.34	0.18	0.13	0.34	0.18	0.13	0.34	0.18	0.13
0.25	0.09	0.08	0.05	0.09	0.05	0.04	0.27	0.12	0.08
0.50	0.11	0.09	0.07	0.16	0.12	0.08	0.30	0.16	0.11
1.00	0.14	0.08	0.06	0.23	0.12	0.06	0.33	0.11	0.08
1.50	0.31	0.18	0.12	0.33	0.26	0.16	0.69	0.32	0.24
2.00	0.43	0.15	0.10	0.25	0.20	0.12	1.05	0.70	0.36

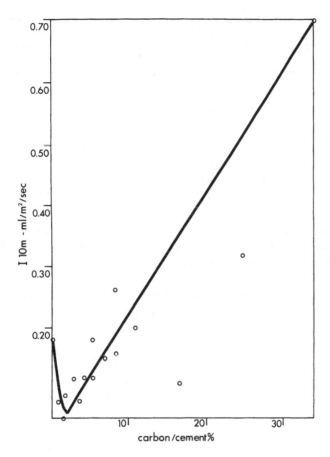

Figure 1.3 I 10 min vs. carbon/cement.

sensitive to curing conditions and, with the co-operation of the precasters, no particular requirements were laid down.

Although hydraulically-pressed concretes are not prone to frost attack, 75 × 75 × 300mm prisms were sawn from the kerbs and immersed in water-filled sealed containers, then placed in an ethylene glycol filled tank. Starting at concrete ages of 29 and 60 days old, they were submitted to a 20°C to −20°C to 20°C cycle every 2 hours. One set was subjected to 100 cycles and the other to 164 cycles. The test is very severe and was selected from the RILEM recommendation. Tests in the then current Standards were not available at that time. Percentage weight losses of averaged specimen pairs are shown in Table 1.13.

The data shows that although the test is very severe, the concretes tested at 60 days old had better resistance than the ones tested at 29 days old with the exception of the control concretes. This improvement might be associated with a pozzolanic effect and/or a change in the elastic properties but the

Table 1.13 Freeze/thaw percentage weight losses for wet-pressed kerbs (number of freeze/thaw cycles in parentheses)

F/C	F1		F2		F3	
age (days)	29 (100)	60 (164)	29 (100)	60 (164)	29 (100)	60 (164)
0	1.3	2.4	1.3	2.4	1.3	2.4
0.25	1.3	0.5	0.8	0.6	3.1	0.9
0.50	3.1	0.4	1.1	0.6	6.9	3.5
1.00	2.0	0.9	1.8	0.9	9.3	3.5
1.50	3.8	0.5	2.6	0.6	14.0	7.1
2.00	4.6	1.9	3.7	1.1	20.0	13.7

evidence is not definite enough to be able to be conclusive. The plot of weight loss against carbon/cement percentage weight shown in Figure 1.4 shows a similar shape to Figure 1.2 for the 60-day-old specimens but not the 29-day old ones.

To study the generally richer (higher cement content) hydraulically-pressed paving flag mixes as well as using top quality low carbon (about

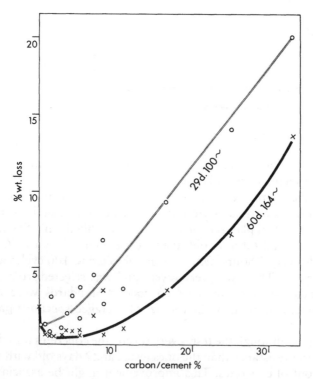

Figure 1.4 Freeze/thaw percentage weight loss vs. carbon/cement in kerbs.

1.0–1.5 per cent m/m PFA) factories 'S' and 'L' were selected to assess ash loadings up to a PFA/cement ratio of 1.0. Details of the mixes are listed in Table 1.14.

The flexural strengths of averages of three tests are shown in Table 1.15, bearing in mind that, at that time, manufacturers would have been unlikely to deliver units with a flexural strength significantly below 7 N/mm^2.

It would seem that the dust-deficient 'L' series benefit more from pore filling than the dustier 'S' series but the improvements are not as marked as the leaner (A/C 5.9) kerb mixes.

Furthermore, it is unlikely on aesthetic grounds that too much of an ash loading would be acceptable. The pozzolanic effect would appear to kick in between 4 and 7 weeks old which indicates that, on site and in place, properties improve more rapidly than the control concretes.

For ordinary wet-cast vibrated concrete both the author's research as well

Table 1.14 Details of wet-pressed paving slab mixes

Ref.	Mix	A	F	Initial water (estimate)
S1	67% granite 10 mm	5.4	0	0.90
S2	33% basalt 5 mm down		0.10	0.93
S3	(very dusty)		0.20	0.96
S4			0.50	1.00
S5			0.75	1.03
S6			1.00	1.05
L1	80% granite 12 mm	5.0	0	0.85
L2	20% natural sand		0.25	0.88
L3			0.50	0.92
L4			0.75	0.96
L5			1.00	1.00

Table 1.15 Flexural strengths of wet-pressed paving slabs

Ref.	Age				
	7 day	14 day	4 week	7 week	18 week
S1	6.8	6.7	7.3		
S2	6.3	6.6	7.3		
S3	6.8	7.3	7.0		
S4	6.7	6.4	6.7		
S5	6.1	6.7	7.5		
S6	5.8	6.0	7.0		
L1	6.6	6.2		7.9	9.2
L2	6.8	8.4		8.1	11.1
L3	6.9	8.3		9.3	9.1
L4	8.7	7.3		10.0	10.6
L5	7.0	8.0		10.0	10.6

as the more recent work on pozzolanic effect by Owens and Buttler[18] have shown that elevated temperature curing up to 80°C with an optimum PFA/cement loading of 0.3 produced very good strength results. This also indicates that the successful use of pozzolanas in Italy and Greece was not only a function of the local availability of ash but the combination with a hot climate.

Initial Surface Absorption and Water Absorption tests were conducted on pairs of flag samples with the results being listed in Table 1.16.

The data shows that the fines deficient 'L' concrete benefit most from the ash addition whereas the benefit for the 'S' mixes is only marked up to an ash loading of about 0.5. The water absorption shows that the cut texture does not fare so well as the surface with increasing ash loadings. This could be associated with the sawing of specimens breaking up the hollow glass PFA spheres and allowing water easier ingress.

Surface drying shrinkage measurements were also made at intervals of 44, 88, 132 and 176 hours of drying starting at about 28 days old. The results were observed to be somewhat random and, although reported in the first edition, are not tabulated here. From the foregoing data, three conclusions can be drawn for wet-pressed hydraulically-manufactured products:

1   Fly ash either has a beneficial effect or causes no property change and the effect largely relates to the mix being fines deficient. The additive is also more effective in the leaner mixes used for kerb manufacture than the richer mixes used in paving flag production.
2   The benefits manifest themselves in improved flexural strength, impermeability and frost resistance although the latter is not a known hazard in these products.
3   Optimum benefits are achieved with PFA/cement ratios in the range 0.5–1.0 m/m cement.

Table 1.16 ISAT and absorption results for wet-pressed paving slabs

Ref.	I 10 min	I 30 min	I 1 h	W 10 min	W 30 min	W 1 h	W 24 h
S1	0.13	0.08	0.05	0.9	1.2	1.5	2.9
S2	0.07	0.05	0.03	1.8	2.6	3.2	4.8
S3	0.14	0.09	0.06	2.0	2.7	3.5	5.3
S4	0.17	0.16	0.10	2.0	3.0	3.7	5.4
S5	0.34	0.20	0.14	2.1	3.2	3.9	5.6
S6	0.35	0.16	0.12	2.0	3.0	3.7	5.6
L1	0.06	0.04	0.03	1.4	1.9	2.5	3.5
L2	0.02	0.02	0.01	1.0	1.7	2.0	3.1
L3	0.04	0.03	0.03	0.7	1.5	1.6	1.7
L4	0.04	0.03	0.02	1.2	2.1	2.8	3.5
L5	0.03	0.02	0.02	0.5	1.0	1.6	1.8

The kerbs used in the first hydraulically-pressed tests were sawn into thirds and placed outdoors on a concrete base and allowed to weather. Two of the high PFA-loaded F3 kerbs split into halves with crystals of ettringite calcium alumino-sulfate visible on the broken face. This was put down to long-term sulfate expansion but of an internally-working mechanism and not from an outside source. The sulfate contribution from the ash plus that from the cement would have given a total SO_3 content of the order of 6 per cent m/m.

Tamped products

An investigation of a similar nature into Kango Hammer-compacted kerbs was made with the kerbs being manufactured in a precast concrete works using the same three ashes described in Tables 1.8 and 1.9. The mix design, all parts m/m, was:

1   Rapid-hardening Portland cement (modern description CEM I/52.5N).
2   10mm granite single size.
4   Natural sharp clean sand 3mm down.

The ash/cement ratios with associated total water ratios were:

Ash/cement	0	0.25	0.50	1.00	1.50	2.00
W/C(T)	0.35	0.39	0.43	0.50	0.58	0.65

The observed and cost-corrected (in brackets) 14-day-old flexural strengths are shown in Table 1.17 in the style of Table 1.11.

It may be seen from Figure 1.5 that there is a similar trend to that shown in Figure 1.2 for wet-pressed kerbs but the spread of data is much wider. This is probably due to the manual intensiveness of the compaction since nominally identical amounts of work would not have been put into each kerb.

These products appear to be much more amenable to the carbon content and can accept loadings up to 10 per cent m/m cement in the ash. The works also commented that with all ashes arris sharpness was improved

Table 1.17 Observed and cost-corrected 14 day-old flexural strengths of pneumatically-tamped kerbs

F/C	F1	F2	F3
0	3.8	3.8	3.8
0.25	4.8(4.9)	4.1(4.2)	4.7(4.8)
0.50	5.0(5.3)	2.7(2.8)	4.2(4.4)
1.00	4.6(5.1)	3.9(4.3)	3.7(4.0)
1.50	4.0(4.6)	4.1(4.7)	2.6(3.0)
2.00	3.6(4.3)	3.1(3.7)	1.9(2.2)

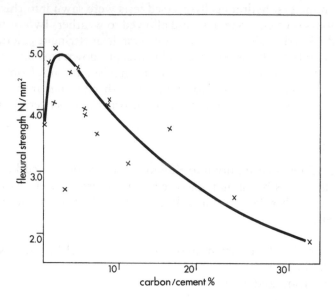

Figure 1.5 14-day strengths of tamped kerbs vs. carbon/cement.

as was the surface finish. The outrider result of 2.7 N/mm² might have been triggered due to trace chloride as discussed earlier in the admixtures section.

Initial Surface Absorption and freeze/thaw tests as for the wet-pressed kerbs were also undertaken but the freeze/thaw tests only ran for 48 cycles from 60–64 days old. The results are listed in Table 1.18.

The data mirrors those in Tables 1.12 and 1.13 but what is more interesting is the approximate linear relationship between the 10 minute Initial Absorption figure and the weight loss. The freeze/thaw test, as stated earlier, is very severe and does not reflect practical weathering. The reason supporting this statement is that third sections of all the kerbs were placed on an external concrete raft as for the wet-pressed kerbs and observed at intervals over a 10-year period. Some edge attrition took place but there was no evidence of general frost damage nor internal sulfate attack. The lesson learnt from this is not to draw conclusions from test observations if the data does not simulate what occurs practically.

The main conclusions drawn from this work on tamped kerbs are:

1   PFA with a range of carbon contents up to 10 per cent m/m ash may be used advantageously.
2   The optimum PFA/cement loading range was 0.5–1.0.
3   Arris sharpness and surface appearance were improved at all loadings.
4   High carbon content ashes at high loadings detrimentally affect the strength.

Table 1.18 ISAT and freeze/thaw percentage weight losses in tamped kerb

F/C	F1				F2				F3			
	I 10 min	I 30 min	I 60 min	Loss	I 10 min	I 30 min	I 60 min	Loss	I 10 min	I 30 min	I 60 min	Loss
0	0.15	0.10	0.07	2.9	0.15	0.10	0.07	2.9	0.15	0.10	0.07	2.9
0.25	0.11	0.08	0.06	3.7	0.32	0.22	0.11	25.1	0.13	0.11	0.12	2.0
0.50	0.21	0.14	0.01	4.2	0.51	0.34	0.24	54.3	0.12	0.09	0.06	3.8
1.00	0.12	0.12	0.04	8.7	0.22	0.14	0.11	13.4	0.09	0.01	0.04	3.4
1.50	0.60	0.48	0.24	8.0	0.24	0.16	0.11	13.4	0.33	0.27	0.23	4.4
2.00	0.08	0.08	0.04	13.0	0.20	0.10	0.08	33.6	0.25	–	0.24	7.4

Extruded roofing tiles

Extruded roof tile production was also examined but, with the restrictions on a mass production process, the factory would only permit one ash to be examined and F2 was selected from Table 1.9. The works mix was a standard 3/1 m/m natural sand with Ordinary Portland cement with a sand grading:

Sieve (mm)	1.18	0.6	0.3	0.15
% passing	100	95	50	5

The following ash loadings and water contents were used:

PFA/cement	0	0.25	0.37	0.50	0.63	0.75
W/C(T)	0.28	0.32	0.34	0.36	0.38	0.40

Observed and cost-corrected (relative to control tiles) are shown in Table 1.19. The steam curing that virtually immediately follows extrusion (80rh, 36°C) did not help the 24-hour strengths for PFA/cement contents above 0.37. This, for F2 ash, equates to about 2 per cent carbon m/m cement. These lower strengths caught up between 7 and 11 days old which might have been due to the pozzolanic effect. However, it needs to be borne in mind that tiles are stacked within 24 hours on extrusion as a rule and the lower strengths at the higher ash loadings would result in too much breakage and thus be unacceptable. Figure 1.6 plots carbon content against 24-hour flexural strength where the same behaviour can be observed as for Figures 1.2 and 1.4. There were not enough ashes tested to draw very strong conclusions but the inference is that the improvements arose through void filling and better workability. The tiles in question were also pigmented to give a red-coloured simulated clay tile appearance but the appearance for carbon/cement contents above about 2 per cent was not attractive.

Initial Surface Absorption tests were carried out and the results are shown in Table 1.20. At all concentrations the values improved over the control

Table 1.19 Observed and cost-corrected relative flexural strengths in extruded roofing tiles

F/C	24 h Obs. (Costed)	7 day Obs. (Costed)	11 day Obs. (Costed)	Carbon/cement %
0.25	1.07(1.10)	1.02(1.05)	1.08(1.11)	1.4
0.37	1.06(1.10)	0.87(0.91)	1.00(1.04)	2.0
0.50	0.83(0.88)	0.87(0.92)	0.95(1.00)	2.8
0.63	0.83(0.90)	0.97(0.95)	0.95(1.00)	3.5
0.75	0.69(0.75)	0.86(0.94)	1.10 + (Machine limit)	4.1

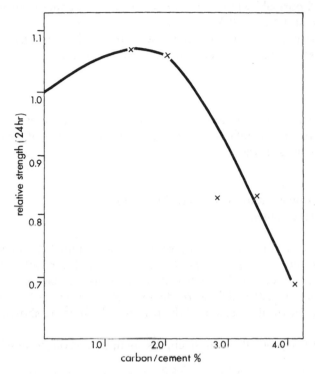

Figure 1.6 24-hour-old relative flexural strengths of roofing tiles.

figures even though the strengths decreased above a F2 loading of 0.37 PFA/cement. This indicates that pore filling and strength are not all that closely related and that individual ash performances would need assessment specific to the ash and the works conditions.

The following conclusions apply to these roofing tiles:

1   Provided the colour effect is acceptable, PFA addition improves the flexural strength.
2   Irrespective of the continuing permeability improvement, the required

Table 1.20 ISAT results for extruded roofing tiles

F/C	I 10 min	I 30 min	I 1 h	I 24 h
0	0.58	0.31	0.17	0.14
0.25	0.40	0.19	0.12	†
0.37	0.10	0.09	0.06	0.01
0.50	0.26	0.15	0.09	†
0.63	0.15	0.11	0.07	†
0.75	0.23	0.19	0.11	†

† Below 0.01, which was the apparatus' minimum sensitivity.

strength at 24 hours old will restrict the maximum ash content; in this case, it was about 0.4 m/m PFA/cement.

3   Being a mortar mix, careful continuing scrutiny of the sand grading would be necessary and the ash content adjusted accordingly.

Wet-cast vibrated concrete

The three ashes were also examined in vibrated wet-cast concretes, the samples being laboratory-manufactured prisms of size $100 \times 100 \times 500$ mm. The mix in parts by weight was:

4   10mm flint gravel.
2   natural sand (equivalent to the modern M grade).
1   Ordinary Portland cement.

The W/C(T) was made up of 0.5 on the cement weight plus 0.15 of the PFA to model on in situ work by others. The prisms would normally have been demoulded at 1–3 days old which was the case for all except those with ash loading above 1.0 which were demoulded between 1–2 weeks old. All prisms were flexurally tested and the results in N/mm^2 are shown in Table 1.21.

A plot of the 14-day-old flexural strengths against carbon content is shown in Figure 1.7.

Initial Surface Absorption Tests were carried out on half-prisms at 28–35 days old and the results are shown in Table 1.22.

Although there is quite a spread of data, the main thing that comes through is that as far as wet-cast vibrated concrete is concerned, under the stated conditions of test any ash addition is detrimental to strength. This does not preclude consideration being given to using ash in wet-cast concrete cured at elevated temperatures. Work by Owens and Buttler[19] and many others indicate that this is the field in which PFA has potential for accelerated (by heat) cured production. Furthermore, high temperature high pressure autoclave curing, as currently used for aerated blocks, is a field worthy of further investigation. Autoclave conventional density concrete

Table 1.21 14 day flexural strengths of vibrated concrete prisms

F/C	F1	F2	F3
0	4.1	4.1	4.1
0.25	3.2	3.7	1.4
0.50	2.6	1.6	1.0
1.00	3.0	1.5	0.8
1.50	3.4	1.4	0.9
2.00	0.4	0.8	0.9

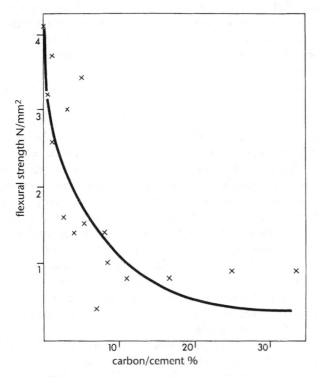

Figure 1.7 14-day flexural strengths of vibrated concrete prisms.

might possibly suit mass production processes such as railway sleepers or tunnel segments.

Other precast concrete processes

If ever spun concrete comes back into vogue, the 1960s work by the Chicago Fly Ash Company showed that the incorporation of ash inhibited the

Table 1.22 ISAT results for vibrated prisms

F/C	F1			F2			F3		
	I 10 min	*I 30 min*	*I 1 h*	*I 10 min*	*I 30 min*	*I 1 h*	*I 10 min*	*I 30 min*	*I 1 h*
0	0.60	0.25	0.14	0.60	0.25	0.14	0.60	0.25	0.14
0.25	0.09	0.09	0.04	0.72	0.31	0.12	0.35	0.30	0.26
0.50	0.04	0.02	0.02	1.00	0.60	0.40	2.25	2.00	1.65
1.00	0.09	0.04	0.03	0.85	0.56	0.45	0.93	0.71	0.42
1.50	0.12	0.06	0.05	0.77	0.74	0.39	1.25	0.72	0.68
2.00	0.11	0.06	0.04	1.50	1.00	0.85	3.25	3.50	2.00

segregation tendency of the various ingredients. The SGs of the aggregates govern what tends to go to the outermost area but a different mechanism based upon particle size governs cement which tends to separate to the inner radius. This is inhibited when ash is present and the ash promotes a more homogenous matrix in the pipe or lighting column section.

Combined vibration and pressure systems used in some pipe processes might well benefit from the incorporation of PFA as the mixes are typically a coarse crushed rock with a natural sand. The mechanisms of pore filling, improved workability and surface finish would be advantages. The author does not know if PFA is being currently used in any of these processes.

Concrete block production (other than autoclaved aerated concrete which already uses PFA as a main ingredient) could also be a candidate. However, concrete blocks are deliberately made to be partially compacted products and the incorporation of PFA might add unwanted weight. This, in turn, would detract from the impact of sound absorption in beam and block flooring. Ash presence might also interfere with the bond performance towards renders and plasters. It is envisaged that the main advantage of PFA in concrete block production would be in its contribution towards the green strength. In view of the possible disadvantages in the aforementioned other respects it is not thought that the high levels of PFA tolerable for wet-pressed and other products would be acceptable.

The roller/rotation method of pipe production might well find PFA addition an advantage in improving workability and void filling. There are possibly quite a few other precast processes where, if PFA is not being used at present, it could be a candidate. As emphasised earlier, all such applications need to be thoroughly assessed under the conditions specific to the actual works process and materials used.

1.5.2 GGBS

Slag is a waste or by-product from several different types of metal ore smelting processes, with iron being the common production source in the UK. Most of the energy used is in the smelting process and grinding the slag down to a fine active material does not consume much energy. The ground granules form an off-white coloured powder mainly composed of calcium alumino silicate. By itself it has little or no cementitious property but, in the presence of alkaline material, it becomes activated and acts as a slow-setting cement in its own right.

At the time of the first edition there was no significant application of GGBS in the precast industry. The material has to come from steel works which are not as widespread as cement factories in the UK. This results in GGBS being not so widely available as PFA without, in many cases, expensive delivery costs.

However, where GGBS supersedes PFA in performance is in the long-term pore blocking process[20] that results in a concrete with superior chloride and sulfate resistance. The large precast (on site) products constituting most of the Second Severn Crossing were made of a GGBS concrete. In order to achieve the optimum performance, the additive is used as a 50–70 per cent cement replacement. The downside of this is a lower early strength but, against this is the advantage which is a reduced early exothermic reaction, a useful attribute for some of the massive concrete structures that are made.

What is not understood is why GGBS was not in more widespread use well before the deployment of PFA. GGBS concretes were used quite commonly in the 1940s manufacture of the invasion barges, many of which can be still seen on the beach at Arromanche with protruding reinforcement still in good condition where it is exposed outside the concrete matrix. Going back further still, it is thought that the Dutch offshore structures built in the late nineteenth century, still in very good condition, were made of a GGBS concrete.

Apart from large precast concrete products destined for marine or estuarine or similar chloride and sulfate exposure combined with its low heat property, the potential for GGBS in the industry is not great apparently. One application was its use at lower percentages than mentioned above in architectural products such as white concrete and light coloured cast stone. The light colour of GGBS makes it much more acceptable than PFA in this respect. The main reason for using GGBS was to inhibit lime bloom in vibrated wet-cast products. Lime bloom would not generally be a problem in cast stone because of the common use of an integral water repellent admixture.

1.5.3 MS

Microsilica (MS) (silica fume) is a by-product from the production of silicon and ferrosilicon and it is called fume because, in the process, the extremely fine silica is emitted as a smoke and has to be collected by specialist means. It is manufactured in countries where relatively cheap hydro-electric power is produced as high current consumption electric arc furnaces are used. The light grey-coloured powder is so fine that it is commonly sold and used as a 50/50 m/m slurry in water which is held in an agitated tank in the works. It is used at a 3–7 per cent cement replacement as it is highly reactive with the free lime and other alkaline ingredients.

It produces much higher early and later strengths than any other additive known at present and, in combination with HRWR admixture, can easily produce concrete with 28-day strength in excess of $100N/mm^2$. The concrete produced is also virtually bleed-free which indicates that extra care is necessary as regards moisture curing. It is an imported additive from countries such as Norway and Canada which have relatively cheap hydro-electricity, high temperature electric arc furnaces being required to reduce silica to silicon. Its high price has to be weighed against its low concentration of about 5 per cent m/m cement compared to the typical 30 per cent for PFA and 70 per cent for GGBS, coupled to its physical and chemical attributes for concrete.

The use of MS is relatively new with one of its earliest applications being in the Norwegian Blindtarmen Tunnel in 1952. For in-depth reading, see the Concrete Society's publication[21] and the relevant sections of Ramachandran's handbook.[22]

1.5.4 *Other additives*

Other research is currently being undertaken into 'new' by-products such as metakaolin, ferrosilicate slag and several others. Provided that each is thoroughly assessed in the operation specific to the precaster, then no objection should be raised. However, a spate of newcomers can be expected as the concrete industry is commonly looked upon as a user of surplus materials. Under pressure of sustainability, the whole industry can be expected to find new solutions. The other side of the coin is that conventional concrete has been used and will continue to be used for encapsulation of unwanted materials including radioactive waste.

1.6 Pigments

1.6.1 *General*

A pigment can be defined as a fine dry powder, slurry or aqueous suspension of that powder, inert to all other ingredients, which is intended to impart a specific colour to the concrete. The word 'intended' forms the basis of this section which concentrates on the practical deployment of pigment application.

Pigments come in a variety of particle shapes and sizes but they all have one thing in common and that is that their particle size is 10 to more than 100 times finer than cement. The mechanism of colouring is the smearing of these particles over the cement and fine aggregate fractions with some effect on the coarser aggregate particles. Obviously the colour of the cement used and the fine aggregate play important roles in determining the colour of the final product and need to be selected with care.

Other, but minor, variables associated with colour are discussed later but one general aspect worth mentioning is fading. There cannot be any such thing as fading as all pigments, with the exception of some of the carbon blacks, are inert. What is defined as fading is the leaching of lime bloom from the cement followed by its atmospheric carbonation masking the smeared particulates. Loss of staining power can also occur on poor quality open-textured concrete when pigment particles can be leached out of the matrix. Both mechanisms, lime bloom and leaching, point to the advantage of incorporating a water repellent in the mix. The viability of using a pigment without this type of admixture is questionable. The chemical fading mentioned earlier of carbon based on, say, lampblack, occurs due to oxidation and is accelerated at higher temperatures in the range 40–60°C which a dark-coloured concrete will promote in hot weather under sunlight.

1.6.2 Pigment types

The most common pigments are the oxides and hydroxides of iron giving the yellow, red, brown and black colours. Oxides and hydroxides of chromium give the greens, carbon gives blacks and the copper complexes of phthalocyanines give greens and blues. The blue colour can also be obtained from cobalt oxide but is a very expensive pigment that needs to be costed on a day-to-day basis on the precious metals market. Titanium oxide is a white pigment and can only be used to best effect in white aggregate and cement concrete. Most of the pigments are at least 90 per cent pure, the remaining content being of innocuous material except for cobalt oxide which can have traces of lead and zinc, making the concrete take on a dark grey slate-like colour with weathering.

The organic phthalocyanines are extremely hydrophobic and very fine and need to be used extended with hydrophilic powder in dry or slurry form to promote accuracy in dispensing. The blue form is alkaline stable but tends to revert to the green form when the surface becomes acidic. This colour change would occur on acid etching or on natural weathering, the latter effect rendering a façade very patchy depending upon elevation and sheltering. This means that blue concretes should be avoided but, if necessary, cobalt oxide pigment should be used.

Carbon black does not describe a pigment as the carbon can be in the form of a highly angular very permeable lamp black base up to a rounded relatively impermeable particle as used in vehicle tyre manufacture. The source should always be ascertained and, like any other ingredient, its effectiveness assessed. Whichever form of carbon pigment is used, it has the tendency to give the mix a lower workability appearance than actually obtains and therefore care should be taken in using a visual assessment.

Most of the iron oxides and hydroxides are nominally rounded in shape. The exception is the yellow form which is dendritic (fir tree shape) and thus demands more water for a given workability than the other iron oxides.

1.6.3 Physical properties

Rather than compare specific surfaces as one might well do for cement, it is best to compare effective particle diameters as this does not take into account the variations in specific gravity pigment to pigment. As there are several different varieties of each available colour, Table 1.23 presents a few commonly used grades. The listed bulk densities emphasise the danger of volume batching but also how different grades of the same colour have different properties. Water absorption data are also given as these are of interest in

Table 1.23 Pigment particle properties

Pigment	Particle size (μm)	Bulk density (kg/m³)	Absorption (ml/100 g)
Red iron oxide	0.1	900	35
Red iron oxide	0.7	1500	20
Yellow iron oxide†	0.2 × 0.3	800	50
Yellow iron oxide†	0.2 × 0.8	500	90
Black iron oxide	0.3	1100	33
Brown iron oxide‡	0.3–0.6	900	50
Brown iron oxide‡	0.1–0.2	1000	30
White titanium oxide	0.2	700	24
Green chromic oxide	0.3	1200	19
Carbon black	0.01	500	100
Green or blue phthalocyanine	0.01	500	N/A††

† The yellow iron oxide pigments are needle-shaped particles and this is why two particle dimensions are given; it can also be seen that they have a higher water demand than the other iron oxide pigments.
‡ Brown iron oxides have a larger range of particle sizes in them than the others.
†† The 'N/A' for the water absorption of the phthalocyanine pigments is because they are hydrophobic in their undiluted form and will not absorb water.

both determining mix water demand as well as in the making up of slurries and/or suspensions.

It is interesting to compare the particle size column figures with that pertaining to Portland cement where the figure would be of the order of 5 microns.

1.6.4 Dispensing and concentrations

The warning about volume batching is repeated as all recommendations are on a m/m cement basis and there is a large spread of densities even for the iron oxide-based ones. When dispensing, domestic-type scales are generally accurate enough for weight or volume-batched mixes provided that the volume/weight ratio for cement is known. Proprietary equipment is available for either weigh batching or volume batching, bearing in mind that slurries and suspensions need to be kept agitated and protected from frost.

Admixtures such as lignosulfonates, carboxylic acids and similar can cause problems in two respects. First, they are wetting agents and will promote lime from the cement leaching and mask the colour. Second, and more importantly, integral water repellent admixtures have already been recommended for coloured concrete and the presence of a hydrophilic agent will counteract and even nullify the water repellent effect. Fortunately most of the water repellent admixture application is in earth-moist products such as reinforced cast stone and reconstructed masonry units where a workability admixture would probably have little effect. The problem arises for vibrated wet-cast coloured concretes where either a special workability aid would be needed or none used at all and workability promoted by the use of selected aggregates.

Concerning pigment concentration on a cement m/m basis, iron oxides are typically used in the 3–6 per cent range but, say, 1 per cent could be used for a tinting and 10 per cent for a very deep colour. Titania would normally be used with white cement and aggregate at a concentration of 1–3 per cent. It has a very high reflective power and when used on in situ steps made with white cement and white calcined flint for a Cement & Concrete Association Open Day had to be washed down with sooty water as it was optically painful to look at in bright sunlight.

Carbon blacks and phthalocyanines are used in the range 0.1–1.0 per cent as suspensions or inert powder dilutions, the concentration being on the pigment weight, not the extension weight. The extension dilution typically lies in the range 1/1–1/10. This dilution needs to be accurately known so that the dispenser or scales can be accurately set to give the required dosage. When suspensions are used, the water dispenser needs to be set to account for the extra water. The earlier warnings about the optical effect of carbon on workability appearance and the extra water demand of the yellow iron oxide pigments should be heeded.

1.6.5 Assessments

There are two aspects to this, first, the pigment and, second, the finished product. They are discussed in the following short sub-sections.

Pigments

As with aggregates, cements, additives and admixtures approved samples should be kept in airtight documented labelled containers. An approved source of material does not necessarily guarantee that there will be consistency in deliveries. It is to be hoped that the supplier will furnish a certificate of conformity or similar documentation to give the precaster a degree of assurance concerning consistency.

Reference is made to the Concrete Society's publication[23] which covers both Standard tests as well as the practical side of application. The main aim of a pigment test is to assess its smearing power on cement. In this simple test a dry mixture of the control pigment and cement weighing 0.5 to 2.0g is placed by the side of a similar mixture of the test blend resting on a white sheet of paper. A square-ended spatula is pressed into the boundary between these two heaps and run for 10–30mm along the paper and viewed with the naked eye or low-powered magnifying glass. This gives two advantages:

1   The colours can be directly compared.
2   If there are unground pigment particles, they will be ground by the spatula pressure and seen as separate smears of colour.

The test can also be undertaken with the pigment alone and can show separate smears with some carbon blacks. This can result in carbon black-pigmented concrete becoming darker than designed due to the mechanical effect of the mixer and scraper blades. Particles of carbon in PFA can also be viewed in the smear test. One special application of the test is where the precaster uses blended cements which are either supplied as such or are prepared in the works. Such blended cements, if also prepared with an integral water repellent, can be kept in bins for long periods without caking.

Products

As with any concrete, architectural or otherwise, the tendering of 100–300mm square samples is generally misleading and does not reflect the variations that will occur in practice on full-sized units even with the best control of all features that one can muster. It is only honest to make a full-sized mock-up of the product, incorporating every feature of the way it will be made. A pocket-sized sample can rarely, if ever, take on board features of geometry, reinforcement, fittings, fixings and insulation sandwiches that the full model can. This, coupled with the difference in materials, workmanship, curing and storage, should generally lead to being able to offer a range of samples exemplifying these variables. These models should be on full-time view both in the works and on site and be targets for both the producer and specifier. After a few months to years, depending upon the elevation and aspect, most precast façades lose these variations and adopt a sameness.

1.6.6 Practicalities

Blending pigments

Tests in the laboratory and in the works using powder dispersers showed that where a works had a large usage of pigment (as distinct from smaller bespoke contracts), advantages could be taken that would soon offwrite the capital costs. These were found to be the following:

1   The tinting/smearing power of the pigment was improved up to double as the powders (cement + pigment + metallic stearate) became intimately blended in the cyclone effect. This, taking the red iron oxide as an example, meant that a 3 per cent powder disperser loading gave the same smear test and concrete colour as 5 per cent added to the mixer.
2   The industrial powder dispersers have the facility to allow the blend to be blown into a holding hopper, thus lending itself to an automated process.
3   The co-blending of the water repellent at the same time as the pigment allows the cement to be held without caking for over a year.
4   An alternative way to blend pigments with cement is the rotary ball mill

which is not only very noisy but can take up to 30 minutes to achieve a uniform blend compared to a powder disperser taking only 1–5 minutes and is relatively quiet. A ball mill will also grind both the cement and the pigment, finer resulting in faster setting and changes in colour.

Manufacture

There are a number of guidelines to follow:

1 Moulds should be of good quality and finish. High gloss paints, plastics moulds and linings should be avoided and matt surfaces are best to avoid hydration staining.
2 Mould release agents should be water-in-oil emulsion creams or chemical release agents. Mineral oils and oil-in-water emulsions promote streaking and staining. Whichever release agent is used, generosity should be avoided.
3 Pan-type or rotary blade mixers are preferable to tilting drum, with mixer and scraper blades accurately set and mixers cleaned thoroughly at the end of each shift and at mix changes.
4 Powder pigments should be weigh-batched and slurries and suspensions volume batched with pigment weights determined. Powders should be added with the cement and slurries and suspensions at the same time as the addition of the mixing water.
5 All other ingredients should be weigh-batched. If volume batching is used, the aggregate and cement bulk densities should be known as well as aggregate moisture contents so that mix uniformity can be achieved by adding the correct amount of water to each batch.
6 Compaction should be as effective as possible.
7 Wooden or acrylic or similar finishing tools will generally be found the best. If metal tools are used, they should be of stainless steel for finishing lighter-coloured concretes.
8 Consistent curing conditions are important and concrete should not be subjected to extremes. Sometimes the use of a membrane curing agent might be found advantageous but the use of covers of polythene or similar is not recommended as this can cause unwanted condensation and staining.

Table 1.24 gives recommendations on materials, pigments and the effects of etching or autoclaving.

1.6.7 Practical considerations

These include the following:

(a) Lime bloom or carbonation, often misnamed as efflorescence, has already been mentioned and is one of the main aesthetic problems

Table 1.24 Selection of materials for variable colours in concrete

Colour Required	Coarse Aggregate	Fine Aggregate	Cement	Pigment	Acid (HCl)	Autoclaving
For smooth-finished concretes						
White	White	White	White	2% TiO$_2$	None	None
Pink	Pink or white	White	White	1% Iron oxide	Darkens	Lightens
Red	Red or pink	Red	OPC	4% Iron oxide	None	Lightens
Black	Dark grey	Dark grey	OPC	5% Iron oxide	Lightens	Lightens
For exposed-aggregate concretes						
Red on white	Red	White	White	2% TiO$_2$	None	None
Black on green	Dark grey	White or cream	White	4% Cr$_2$O$_3$	None	Lightens
White on black	White	Dark grey	OPC	5% Iron oxide	None	Lightens

with concrete. It is a particular drawback with dark coloured concrete as can be seen in the carbon black coloured walling blocks shown in Figure 1.8. The incorporation of a metallic stearate powder in moist mix design products or srearic acid powder in wet-cast vibrated products is advised. For all products the concentration should lie in the range 1–2 per cent m/m cement with the loading required determined by works trials. As with all admixtures the usual full attention should be paid to

Figure 1.8 Lime bloom fading and staining on wet-cast masonry units

attaining as full compaction as possible coupled with appropriate curing conditions.

(b) Mix appearance when carbon blacks are used can be misleading as the surface looks viscous and oily and there is a tendency to mistake this for low workability and add unnecessary extra water. Once the water needed for the design workability has been determined, this should be adhered to for all mixes.

(c) Autoclaving generally has a lightening effect on all pigment colours but it needs to be ascertained which pigments are stable at the conditions of curing. Some carbons can oxydize and the phthalocyanines need to be specifically assessed. Where autoclaving does cause a fading, the pigment loading needs to be increased to attain the design level simulating normally-cured concrete.

(d) The usual precautions should be taken when handling pigments especially as they easily become air-borne irritants due to their small particle sizes. Special care is necessary for chromic oxide with operatives being first cleared for allergic reaction and skin sensitivity. Furthermore, washings from chromic oxide pigmented concrete should never be allowed to enter mains drains or water courses as chromium is extremely poisonous to plant and fish. Advice on collection and disposal of all effluents should always be obtained.

(e) Acid etching, still deployed by many manufacturers and contractors, is not recommended. Apart from any benefits being generally short term, it promotes further lime bloom as the calcium chloride left in the pores is hydrophilic and attracts more leaching of lime. It also will turn the blue form of phthalocyanine into the green form.

2 Reinforcement, prestressing, hardware

In the first edition of this book all these items were grouped under the heading of 'Materials'. It was decided to place them in a stand-alone chapter in order to give each item specific attention. Since this book is specifically written for the precast concrete industry, highlights from the first edition as well as those from the more recent book by the author[24] are discussed.

2.1 Steel reinforcement

Carbon and stainless steel reinforcing bars from accredited sources are certified under the CARES[25] system where the origin, type and details of each bar are coded into the surface. CARES offer certification schemes for both types of steel as well as a special scheme for precast concrete products. Such steels should be purchased with a Certificate of Conformity or similar certification showing compliance with the Standard.[26] Stainless steels should be to BS6477[27] which lists six stainless steels for use in concrete out of the total of 60 listed in the European Standard.[28] Again, all consignments should be accompanied by a Certificate of Conformity or similarly acceptable documentation. Reinforcing steel is used in precast concrete products for one or more of the following three reasons:

1   To enable the product to take loads mainly in the tensile region of the product where, otherwise, cracking could lead to collapse/failure. In effect, the steel takes over the tensile effects due to concrete/steel bond being sufficient enough to transfer the load.

2   To enable the product to withstand handling where a region concerned would otherwise be in compression in the works.
3   To enable a region of the product normally in compression to withstand shrinkage caused by effects other than loading.

Each one of these applications is discussed in the following sub-sections but, before doing so, the essential points about cover to reinforcement need highlighting.

Loading, and in its turn, structural performance, depend upon the cover and this subject was discussed at length in the book on concrete materials.[29] The salient points of that discussion are recalled here. There are two critical factors to cover which are:

1   the amount (distance from surface);
2   the quality (usually taken as permeability) of the covercrete.

It has taken many years for the thinking in the concrete world to move a little away from the concentration on the importance of 'd', the distance of the reinforcement from the surface, to 'k', the permeability of the 'cover-crete'. Specifications rarely mention both d and k and rely on giving a minimum number for d and catering for k by strength, mix design or minimum cement content requirements. The old philosophy on d would have seem to have been that the depth of cover is that which determines when the fatal date of corrosion commencement begins. The rule of thumb was that 20mm of cover gave 20 years, 40mm 40 years, and so on. However, the importance of k is exemplified in several ways:

1   A mortar split ring spacer, illustrated later, used to be in common use. The half rings were kept together by a soft iron wire resting in a rebated groove in the circumference. The rebate was only about 2mm deep, giving the wire about 1mm cover. During nearly 20 years usage known to the author only one case of corrosion of this wire was reported. This, on investigation, was found to be due to the spacer having been used on an oversized bar and caused the wire to rise up to the surface.
2   The concrete barges used for the Normany landings very often had cover as low as 10mm and were beached in the tidal zone. With four decades (at the 1984 approximately report stage) with bars protruding from the broken concrete, corrosion had not travelled into the concrete even though the bars were heavily rusted. Several examples of these barges can be seen on the beach at Arromanches and other landing sites.
3   In the manufacture of ferrocement boats, the carcass cage of reinforcement is designed for zero cover as operatives, working in pairs, push a high cement content mortar towards each other with trowels to fill up the interstices. To prevent rust showing, the concrete is painted.

These examples show that d and k need to be considered together. This aspect is discussed later in Chapter 9 on properties. If d is considered the only factor and the concrete is of a mediocre or poor quality, then carbonation will work its way from the surface, then depassivate the protective oxide layer on the steel and then start corrosion. If chlorides or other aggressive agents are present, then the initiation of corrosion will be accelerated. The choice is in the hands of the designer. If corrosion of steel is the only criterion, then a few millimetres of cover is sufficient provided that concrete with a low k is used. The other factor favouring low cover is that under design load the stress in the tensile zone will result in a given strain. Since this strain is catered for by the steel, then cracking will result. For very deep cover the strain transferrence will also be very deep and the surface cracking observed is likely to be in the form of just one or two cracks with quite wide crack surface widths. The shallower the cover, the larger the number of surface cracks but their surface crack widths will be smaller in comparison. This is why cracking in fibre-reinforced concrete can be almost invisible because of much surface cracking taking place but with too small a crack surface width to be observed.

Cover is also of great importance in specifying fire resistance where the old rule of thumb was that one inch gave one hour's resistance. This is rather an over-simplification in that although d is apparently the most important factor, the type of aggregate is also important. If the aggregate is flint, it will calcine and expand at fire temperatures (i.e. 800–1000°C) leading to fissuring and cracking. Such concrete might well require secondary mesh-type reinforcement in the cover zone to hold the concrete in place to assist in the protection of the steel.

Volcanic aggregate sources, sandstones and limestones generally behave in an inert fashion and do not require secondary reinforcement. However, limestones decarbonate at about 900 °C and turn into quicklime which, as it cools, slakes and falls out of the surface. This reaction is accelerated by the action of fire fighters' water and can be violent for some limestones. The fire authorities tend to take a calmer approach to fighting fires inside concrete frame buildings as they receive warning signs of collapse well in advance of danger. Reinforced units beginning to lose surface aggregate and prestressed unit beginning to hump as the wires/strands lose efficiency in the prestress zone are early warnings.

2.1.1 Loading

The amount, type and configuration of steel reinforcement are the main three points of interest. In the early days of reinforced concrete precast construction, typical steel cross-sectional areas were 1–2 per cent but it is not uncommon nowadays to observe a considerable excess of this. Very often it is difficult and sometimes impossible to see the bottom of a mould due to steel congestion. This places restraints on the precaster and promotes the use of plasticised and, more generally, superplasticised or self-compacting concrete. In order to ensure that the concrete reaches and fills all the in-steel voidages, glass or acrylic observation ports may need to be designed into the mould or non-destructive techniques such as ultrasonic testing deployed. Reference was also made earlier to the importance of the steel/concrete bond and this is promoted in current steels through the bars being contoured. In addition to this mechanical bond, there is also bond due to the alkali-induced surface passivation. Additional bond can also be obtained by allowing the steel to acquire a fine degree of rust corrosion. However, this must not be allowed to extend to deep rust and/or flaking as this will result in poor bond and promotion of corrosion.

Reinforcing cages should be prepared on a jig bespoke for the particular product in order for the required dimensions and tolerances to be achieved. Spacers, discussed in Section 2.4, can be used to achieve this for products such as fence posts where a square or rectangular-shaped ring spacer controls both cover and positions of the steel bars. Bad positioning of the reinforced cage as shown in Figure 2.1 gives the industry a bad name. In Figure 2.1, scaled by the 50mm diameter lens cap, the cover in the nib is negative, i.e. protrudes from the face. The shims also show that dimensional tolerance was not a strong point; whether or not this applied to the unit or the receiving structure or both is not known. Figure 2.2 shows precision placing of protruding reinforcement loops in precast floor planks designed to locate onto the studs of the main frame steel beams whence an in situ joint would be made.

Figure 2.1 Cladding panel with poor reinforcement placement.

Figure 2.2 Precision reinforcement loop placement in floor panels.

Steel reinforcement can also be galvanised, sheradised or coated by other means such an application of zinc-based paint. Galvanising is not only preferred for its permanence but because there are industrial facilities that can take whole cages of reinforcement and hot-dip galvanise them completely. Making a cage from galvanised bars causes problems as welding removes the zinc and, if left untreated, a bi-metallic corrosion cell will be set up. If galvanised steel is welded, the joints should be cleaned and a proprietary zinc paint applied.

Care needs to be exercised with tie wires for the same corrosion reason. Mild steel wires should be used with non-zinc-coated steel and galvanised steel wires with galvanised steel reinforcement. For all types of tie wire, the tie should be turned into the body of the concrete and away from the cover zone.

The surface condition of the steel is an important factor in determining how good the concrete/steel bond is to be. In the CARES system all bars are ribbed with the coding stamped between the ribs. The bond can be further improved by allowing the bars to weather so that a thin film of rust forms on the surface giving micro-texture bonding as distinct from the ribbing macro-texture. This method of promotion requires special control with the steel not being allowed to the point of having loose and/or flaking rust.

2.1.2 Handling

The incorporation of reinforcement to cater for the weak tensile nature of concrete is generally a design aspect relating to service in use. An example was given[30] where precast concrete individual steps for a fire escape were handled by lifting at their ends, resulting in tension in an in-use zone that would be under compression. In the first edition a full account of the events was not given and this will now be rectified.

Extra steel was incorporated into the replacement units as the first batch was damaged during handling. The steps in question were about 16 in number and were intended for a post tensioned (to ground anchor) spiral staircase to access a railway signal box. The cantilever part of each step had top reinforcement to cater for foot loading. However, on arrival on site each was lifted off the transport by two men turning the unit into a simply supported beam. All the units cracked badly at the soffit junction and were scrapped. Another set of units were made with both top and bottom reinforcement in the step and were delivered and handled safely. However, they were prestressed without packing mortar at the interfaces so that trowelled faces were prestressed against moulded faces. The severe pillar cracking that resulted is illustrated in Figures 2.3 and 2.4. The units were scrapped and another 16 units were manufactured as per the second batch and assembled with 10mm mortar joints with the specified step to step winding. By the time the contractors had reached the 15th unit the step was level with the signal box platform and one step was left over due to 15 mortar joints addition to the height. Not only that but the gap between the 15th step and the platform constituted a safety risk.

The whole system was discarded and wooden steps substituted.

It is also not uncommon for units such as relatively long reinforced sills or lintels to be handled in their upside down mode and handling reinforcement could well be an advantage where such a site handling risk is present.

2.1.3 Shrinkage

Duct covers, especially decorative ones with architectural finishes, would have their top surfaces under dead load compression but would be subject to wetting and drying shrinkage cycles. These movements would be relative to the tensile zone which would generally be under relatively constant temperature and high humidity conditions. The top reinforcement would only need to be of a light mesh type or even fibre reinforced.

2.2 Fibre reinforcement

This section is not intended to be exhaustive and highlights the main considerations concerning applications in precast concrete products.

Only in the case of steel fibre reinforced concrete may a tenuous comparison be made with the more conventionally-reinforced steel bar-reinforced concrete. Since glass, polypropylene and other plastics fibres have relatively low tensile strengths compared to steel, the tensile requirements cannot be so effectively transferred to these fibres. Where glass and plastics fibres come into their own is in crack control and/or impact-improving

Figure 2.3 Post-tensioning-induced interface spalling and cracking in bedding mortar-omitted step units.

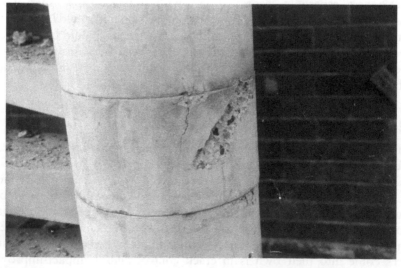

Figure 2.4 Close-up of Figure 2.3 staircase showing butt joints and severe spalling.

properties. These applications are technological developments and rather more advanced than the biblical use of straw in brick manufacture and the more recent use of horse hair in plaster.

European Standards cover virtually all that is required to specify and use steel, polymer and glass fibres, with glass fibres being treated as being more specific to precast concrete rather than in situ concrete. A general Standard covers all fibres with reference concrete being specified[31] and a specific part relating to strength.[32] Steel and polymer fibres are defined with specification and conformity requirements[33,34] and there is a stand-alone Standard specifically for strength retention of glass fibre-reinforced precast concrete.[35]

There are a large number of proprietary fibres available and more can be expected to arrive on the market. Bond is improved in steel fibre systems by having the fibres contoured. Polypropylene fibres are fibrillated to give 'sheaf' type contours. Glass tends to have a chemical bond through reaction with the alkaline parts of the cement but is also woven to produce contour bonding in addition.

As a general summary, all fibre systems act as crack control but only steel systems significantly improve the ultimate behaviour under load.

2.3 Prestressing

A common form of prestressing in the precast industry is the wet-cast (vibrated) pretensioning method where long line production (moulds in line on a long bed) is used for beams, slabs and railway sleepers. The wire and/or strand used needs to comply with the European Standard[36] with each reel being accompanied by a Certificate of Conformity.

As discussed and illustrated in the first edition, products may also be made by the extrusion process where concrete is compacted by, typically, a machine propelling itself along a long line of prestressing wires/strands by the extrusion force from the augers acting on the concrete. Post-tensioning processes, either with or, more commonly, without pretensioning, tend to be used for larger civil engineering applications. One specific mould application is countercasting where the side of one mould mirrors the counter-side of the next mould leading to an accurate positioning of ducts for the prestressing steel. Westway on the M40 coming out of London is one of the early examples of this application.

Prestressing wires and strands are normally delivered on large spools with each spool accompanied by a Certificate of Conformity showing compliance with the EN.[37]

2.4 Spacers

The main purpose of spacers is to retain the cover to the steel reinforcement from the exposed face (visual or exposed to air in a cavity or at unit ends) in the design position. The attainment of this aim in no way absolves the designer or precaster from ensuring that the quality of concrete in that cover zone is fit for purpose. If the quality is poor and allows degradation in the form of carbonation to occur down to the steel with initiation of corrosion, the cover depth will only relate to the fateful date.

Spacers mainly fall into two groups:

1   Wheel-type, being mainly circular or semi-circular in shape and giving virtually equal cover from all spacer sides.
2   Trestle-type, being of a chair or single seating shape giving the design cover from only one edge only.

Wheel-type can be used in any mould position but are not, as a group, generally as robust as trestle-type and might well require more spacers per bar length than for the trestle group. However, wheel-type are often easier to manufacture with piercing and therefore give better homogeneity with the concrete, better fire resistance and a more tortuous path for corrosion-inducing hazards.

At this point more emphasis need to be placed on the use of plastics spacers under cold conditions. Below about 5°C, polythene, the commonly used polymer in spacers, becomes much less elastic and subject to brittle failure, especially the clip-on variety. These need to be stressed to open to fit the steel.

Under the action of vibration and other induced movements most wheel-type spacers can rotate and still maintain the requisite cover. Trestle-type spacers could lose cover and therefore should only be used on mould bases or on shallow sloping mould sides. The earlier mention of cold weather handling leads to another tip that applies to wire clip-held wheel-type spacers as well as trestle spacers that can slide along a bar and that is the judicious use of elastic bands which are quick to instal and are of no corrosion risk.

Spacers made of plastics need at least 30 per cent of their cross-sectional area pierced to allow the concrete to interweave during compaction, thus counteracting the relatively high thermal expansion which is of the order of 10 times that of concrete. The piercing also gives a vastly improved fire performance where, otherwise, the fire would have direct access to steel once the plastics had been destroyed.

The only variation from the discussion in the first edition is that asbestos cement spacers are no longer allowed. How much of an extra risk this may lead to in demolition work needs to be assessed by an expert in this field.

Trestle spacers may also be made of steel with corrosion inhibited by plastics 'thimbles' or 'slippers' either placed or slush-moulded onto the feet. These thimbles need to be carefully examined before deployment to ensure that they are undamaged. The paper by Levitt and Herbert[38] deals with precast and the Concrete Society's publication[39] deals with general concrete applications as well as the British Standards.[40,41]

Considerable research was undertaken in the paper[42] and is briefly summarised here. Several makes of trestle, wheel and the split mortar spacer were placed on a reinforcing bar with three welded protruding plates on it and cast into 75 × 75 × 150mm prisms of a C45-type concrete. The three welded plates on each bar protruded from the trowelled face 5–10mm and were loaded until the concrete was 24 hours old, simulating a heavy reinforcing cage. Thus the load was applied to the two spacers on each bar. Figure 2.5 shows a load frame after 10 years weathering together with three bars shown against the spacer tested with the plates removed. Close-ups of a plastics and a mortar wheel-type are shown in Figures 2.6 and 2.7.

It may be seen that slight corrosion occurred after 10 years weathering in the case of the plastics spacer and that the interlaced concrete was not

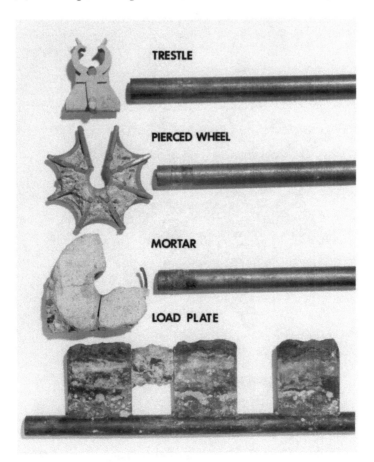

Figure 2.5 Reinforcement frame for loading spacers after ten years.

perfect. A water passage would have been promoted along the rib seen at the 10 o'clock position. This implies first of all that the piercings in plastics spacers must be large enough to allow the maximum aggregate size to nestle comfortably in these voids. Second, the workability must be high enough to permit easy flow of the mix into the piercings.

The steel in the mortar spacer case shows considerable corrosion probably due to moisture access via the spacer split seen about the 6 o'clock position. It would be difficult in concreting to prevent these split mortar spacers rotating because corrosion does not occur when the split is parallel to the weathering face, as may be seen in Figure 2.8.

Figures 2.9, 2.10 and 2.11 show the interweaving of the concrete through the piercings of three wheel-type spacers. Figures 2.9 and 2.10 have soft iron wire holding clips which could be replaced by the use of elastic bands previously mentioned. Figure 2.10 is more correctly described as a half-wheel-type spacer octagonal in shape rather than circular and only gives the

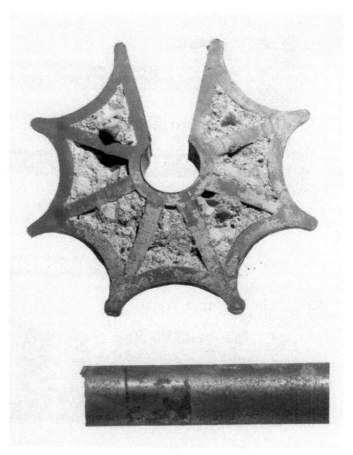

Figure 2.6 Close-up of plastics spacer after 10 years weathering.

Figure 2.7 Close-up of mortar wheel spacer after 10 years weathering.

designed cover over 180°. Figure 2.11 has a sprung plastics entry for the bar and could give problems in very cold weather due to the plastics decrease in elasticity and proneness to become brittle.

Figures 2.12 and 2.13 illustrate the effects of differential thermal expansion between the polythene in an unpierced trestle-type spacer and the surrounding concrete leading to surface spalling. The cruciform shape of the spacer base can be clearly seen and has been known on site to show exposure of the steel reinforcement at a few months old.

Figure 2.8 Mortar wheel spacer with joint parallel to surface after 10 years weathering.

Figure 2.9 Interwoven concrete in pierced wheel spacers.

2.5 Cast-in materials

In addition to spacers, there are other types of hardware used in precast work. The most common of these are cast-in lifting sockets which are integral with the reinforcement in order to dissipate lifting forces into the mass

Figure 2.10 Interwoven concrete in pierced half-wheel spacers.

Figure 2.11 Interwoven concrete in pierced full-wheel spacers.

of the concrete. Sockets and their accompanying bolts are commonly of stainless steel as they will often be exposed until installation. Those made of bronze alloys might possibly need electrical isolation from the steel in order to prevent bimetallic corrosion in certain exposed conditions. Many of the proprietary devices have swivel heads in order to minimise stress when

Figure 2.12 Plastics unpierced spacer inducing spalling due to differential thermal expansion characteristics.

lifting and handling forces are not applied orthogonally in direction. When a unit is to be lifted from two or more lifting points, it is advisable to use a spreader beam from the crane so that the force applied to each lifting point is at right angles to the surface.

Discussed in the first edition was the application of cast-in electrical conduitry and this, together with cast-in heating, can be predicted as growing applications. Figure 2.14 shows this application. The circular socket shape was because the photograph was taken in an overseas company.

With energy conservation and other properties of interest in mind, cast-in temperature sensors and monitoring devices for moisture content are some of the refinements that are worthy of consideration. Humidity sensors in suspended ground floor units might be of particular use for constructions built on soils subject to severe shrinkage which could lead to damage to the property. These sensors could be coupled into the water drainage pipework and allow water from baths and showers to divert onto the ground until optimum conditions could be reinstated.

Although not strictly in the cast-in hardware group, the manufacture of units with external metal frames has had an application in the machine tool industry. Such units are not only cheaper in material and production costs than steel but have low maintenance costs, especially from the point of view of corrosion. Of equal or possibly greater note is that concrete machine tools have good dimensional stability. Once the concrete is matured and

Figure 2.13 Plastics unpierced spacer inducing spalling due to differential thermal
expansion characteristics.

has reached its equilibrium moisture content with its surroundings, it is
extremely slow in response time to changes in the surrounding conditions. A
further added advantage is that if the concrete is damaged (e.g. spalled), it
can easily be repaired at room temperature using well-established polymer
mortar techniques. Only if the edge frame becomes damaged might there be
a need for welding repair.

Cast-in stainless steel dowels in small unreinforced units such as balusters
have been used for many years as they act as locating and fixing references
not subject to loading. In balusters not only may off-cuts of CARES-
approved stainless steel be used but, provided the quality of the steel is
acceptable, lengths cut from stainless steel studding also serve good purpose.
In a natural Portland stone cleaning and repair contract overseen by the
author, some of the dentils were so badly weathered they had to be replaced.
The contractor made a couple of wooden moulds and used the Portland
finish polymer repair mortar as the mix. The mould had a hole at its end
from which protruded the cast-in stainless steel dowel, in this case made
from studding.

All the foregoing sounds tremendously enthusiastic about what one can
put in and around concrete but, and it can be a big 'but', great care needs to
be exercised in ensuring that there is enough space among the medley for

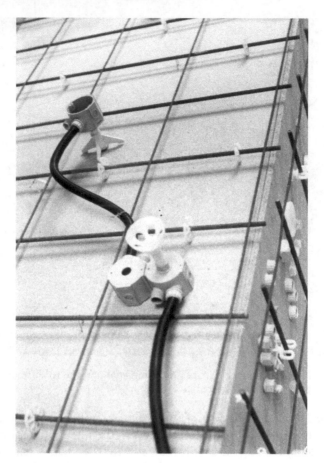

Figure 2.14 Application of cast-in electrical conduitry.

concrete to infiltrate and be thoroughly compacted. With reinforcement (primary and, possibly secondary), prestressing wires and strands, lifting devices, spacers, conduitry, ducting and all the other paraphenalia, it probably pays to produce a full-sized mock-up. This would serve two purposes. First, one could observe any in-works possible problem areas; second, a unit could be cast and assessed for filling and compaction efficacies.

3   Moulds

3.1 General

Chapters 5 and 6 discuss, respectively, labour- and machine-intensive processes. In the former group, moulds are virtually independent of the remaining plant, whereas, in the latter, moulds are either part of the machine or are largely integral with it. Notwithstanding these differences, all moulds have two common requirements:

1   maintaining a required geometry;
2   maintaining the required tolerances.

Both these properties relate to dimensions which are critical for structural and/or contractual and/or architectural reasons.

Structural properties relate to the product being of dimensions suitable for safe stacking, handling, loading and an adequate resistance to the strains imposed on site due to load, movement and temperature and/or moisture effects.

Contractual requirements relate to the ability of the product to fit its assigned location and to the flexibility, adjustability, longevity and strength of fittings and fixing devices.

Architectural needs from moulds generally relate to shape and/or surface finish and could refer to deviations from planeness of a visual surface as well as to the finish.

Tolerances are the overriding consideration in all three of these requirements and a factor that is given too little prominence. When it comes to placing a precast concrete product into its designed location on site, the lesson to be learnt is that it is easier to find tolerance during installation than to lose it. If a product has too much negative tolerance on a designed dimension, this shortfall can be tolerated by the use of packing, sealant, mortar or similar. When there is too much positive tolerance, the designed joint has to be modified with cutting being required or other labour-consuming operations have to be considered.

Precast units such as beams and columns need to locate in specified positions and cannot be given too generous a negative tolerance. This points to the necessity of tighter tolerance than would be considered for a product such as a duct or a block where more licence in building-in is acceptable.

For the vast majority of products this points to having a larger negative tolerance on dimensions than a positive one or even having a zero positive tolerance. Another factor favouring this for many types of mould material is that moulds tend to grow in size with continued use and addition to the positive tolerance can occur.

3.2 Materials

The following sub-sections discuss the most common types of materials used for moulds but do not denigrate in any way the use of other approaches for moulding concrete into its required shape. Provided that robustness and stability are taken into account, cardboard tubes, corrugated boards and the like can have acceptable applications. The choice of material or materials (i.e. composite constructions) is a function of how many uses are required, what sort of unit is being made, what tolerances are required in geometry and surface appearance. Post-contract possible requirements may well justify having a mould too good for the existing work but retaining it in store for future work. Some moulds, especially timber ones used for architectural cladding work, can be manufactured to extra quality and stored after use for possible future extension work to a building. It was a pity that this was not envisaged for one of the Laingwall-clad single-storey buildings where an extra storey height was built at their headquarters (now demolished). This necessitated a new mould to be made and used for about 16 casts. The mould was later stored in the company's depot awaiting possible future use.

3.2.1 Steel

Steel is a common material for many machine-intensive processes such as hydraulically-pressed, packerhead pipe and machine block production and is often selected for labour-intensive production where a large number or uses is required. Figure 3.1 illustrates this latter use for a tilt and turn process. Figures 3.2 and 3.3 show steel moulds being used in dropside productions of tank manufacture. Figure 3.2 illustrates the practice under factory conditions whereas Figure 3.3 takes a more rural approach.

Steel is the first choice of mould materials due to its high strength, abrasion

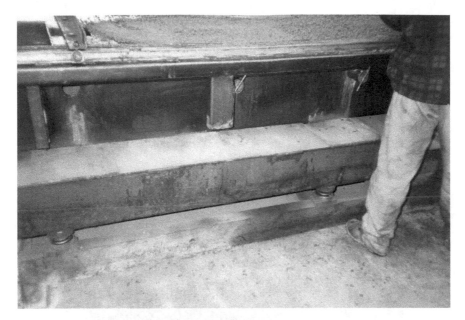

Figure 3.1 Steel mould used in tilt and turn production.

Figure 3.2 Steel moulds used in box unit production.

Figure 3.3 Steel moulds used in box unit production.

resistance and good temperature stability compared to timber and plastics. This is no reason to become complacent as the condition and dimensions need checking on a regular basis and the precaster should include this activity in the maintenance schedule. The lifetime or number of uses from a steel mould will be a function on the type of process and the degree of care exercised in maintenance. In labour-intensive processes, as illustrated in Figures 3.1, 3.2 and 3.3, well over a thousand uses is not untypical. In machine-intensive processes the mould is normally of a more robust nature than for labour-intensive processes and lifetimes could be in terms of several thousands of uses.

3.2.2 Timber

Timber remains one of the most useful materials for bespoke production where a limited number of uses (viz. 5–100) is required. However, timber is available from a variety of trees and none of them is as stable a material as steel. The geometry can change with moisture content, absorption of release agent and ageing and dimensions need regular checking to ensure that tolerances are maintained within design requirements. Mould protection is discussed under 'Treatments' (see Section 3.3) and it cannot be emphasised too strongly that a careful consideration of selection and preparation pays dividends in the long run. At the risk of being challenged by the timber supply industry, the quality of the best available timber (with the possible exception of hardwoods) nowadays does not stand up against the best of the Douglas Fir and Russian Redwoods one purchased in the 1960s and 1970s. Therefore, selecting the best available and storing in conditions such that the timber neither dries out too quickly or too much nor becomes too moist is advisable. All too often deliveries have been observed where warpage occurs necessitating straightening, planing or cutting operations.

Timber is also available as plywood, chipboard, fibreboard and wood wool slabs, each of which can have an application as a mould material as a concreting surface and/or as a composite with other mould materials. Some of the latest developments in plywood manufacture are those being used in timber frame house construction where the alternate sheets have the sheet grain directions at right angles to the sheets on either side. This results in a much stronger composite whose extra cost could be justified either in the extra performance and/or being able to use thinner sections.

It is advisable to have timber moulds prepared by professional carpenters who would qualify generally as pattern makers. After all, whatever shape is the void of the mould will be the shape of the product. Any imperfections in the mould will show as imperfections in the casting. Whether or not mould manufacture is an in-house or externally contracted activity is the choice of the precaster. The same requirements apply.

Where timber or composite moulds are made by external agencies it needs to be remembered that the mould might well need re-fettling from time to time. This means that if the precaster has not got an in-house facility to deal with this then the mould has to go out of commission for (hopefully) a short interval.

Figure 3.4 shows a robust timber mould used for the manufacture of a cladding panel. The uprights at the corners were to support a suspended reinforcement cage so that the bottom-as-cast visual face would not show spacers.

The lifetime of a mould depends upon many factors, the most important being the paint used to protect the surface; this is discussed more fully in a Section 3.3.2. Generally the number of uses will vary with circumstances from 20 to 100. However, timber has the advantage that it can be replaned and refurbished so that corrective measures can be quite economic. When wooden moulds are stored for later usage, they should be kept in medium humidity conditions to promote dimensional stability that could well be sacrificed with large changes in moisture content. Moulds should not be

Figure 3.4 Robust timber mould for cladding panel production.

subject to live loads and the effects of dead load should be minimised wherever possible. If the outsides of the mould have not been painted, a useful attribute for lifetime promotion, they should be treated with a thin film of water in oil emulsion or chemical release agent. This helps to preserve the timber and deter insects. Oil in water emulsions should not be used as they can cause the timber to swell.

3.2.3 Concrete

Concrete moulds came into prominence in the 1950s for the manufacture of tunnel segemnts and are still in use today for the manufacture of these and similar units where very strict tolerances are specified. None of the other mould materials discussed is capable of reaching the tolerances achievable with concrete moulds. Their disadvantage is their weight, leading to the requirement of mechanical handling for both product and for the mould. In addition to the tolerances achievable, not even steel can compete with mature concrete's temperature and moisture stability, resulting in long-term performance and strict dimensional control in production.

As an example of the requirement for some tunnel segments the curved rhomboid-shaped sections have alternating rhomboid edges which mate up to each other with a circumferential tolerance length specified to +/− 0.5mm maximum. The reason for this is that the units are hydraulically rammed into place with the invert unit locking the complete ring. The manufacturer concerned worked to a nominally zero tolerance throughout the contract.

A rich mix design is used for the mould manufacture with a typical cement content of 350–400 kg/m^3 and a maximum W/C(F) of 0.45. For the above example the moulds were made slightly oversize and allowed to mature for several months then ground down to size using a steel template jig. The jig was only used over a strict temperature range being of steel. The coefficients of thermal expansion of concrete and steel are of the same order but the thermal inertia of a few tons concrete mass is low, whereas steel, being generally of a thinner section, has a faster response to changes in temperature.

With proper care and treatment a lifetime for a concrete mould would relate to several thousands of uses.

3.2.4 Plastics

In the form of thermoset plastics such as glass-reinforced polyester (GRP), glass-reinforced epoxide (GRE) or thermoplastics such as polyvinyl chloride (PVC), polystyrene or latex, plastics have a variety of uses with their main advantage over steel and timber being shape. Complex geometries are expensive to produce in other materials. As an example of this, synthetic rubbers were used for cast stone building 1939–1945 blitz damage repair. Acacia leaves and berries at the heads of Corinthian columns, were cast in moulds, peeled off when the cast stone was hard, and pinned to the head. No other material could have been usefully deployed to cast a detail with both positive and negative profiles.

Figure 3.5 illustrates a more common application of a GRP mould in use for the production of an end-stop gulley box unit. GRP moulds and form-work are suitable for products such as garage panels, gulleys, flower pots (especially frustrum cone-shaped units) as well as in situ coffered floor systems. To maintain dimensions, composite construction is generally required. For example, in the case of a panel, not only would edge reinforcement be required but the flat area at the base would need chipboard or similar external reinforcement to counter twist and bowing. Deep GRP or GRE moulds might also need cast-in metal plates to accept clamp-on vibrators. Most moulds in thermoset plastics will require edge reinforcement to resist damage and to maintain dimensions. In situ coffer units will normally be without edge reinforcement and would be supported on falsework with the concrete-contacting edges of the joints being taped or similar. GRP moulds, the most common of the thermoset plastics, can yield 200–1000 uses if maintained well. Damage can be repaired with a cold-setting mix but it is difficult to achieve anywhere near a good matching surface without extensive grinding and polishing work.

Thermoplastics are used either as mould linings or can be vacuum-formed into complex shapes and used as moulds if robust enough or supported

Figure 3.5 GRP mould for an end-gulley unit.

if not. Something like 10–50 uses are possible as aggregate abrasion detrimentally affects the smooth finish generally required.

Both types of plastics have thermal expansion coefficients up to 10 times higher than concrete or steel. Mould linings need to be firmly adhered with total adhesion to the base with the mould being used under as uniform a temperature condition as possible. This is to avoid as much as possible warping or curling of the plastics. The heating effect due to cement exotherm should be minimised by not casting too thick a section and heat acceleration of curing avoided. In all cases full trials of the process are recommended.

3.2.5 Aluminium

The main use is in roofing tile production where aluminium (or more strictly, aluminium alloy) is used for the pallets that travel under the extruded mortar ribbon. Many thousands of uses can be obtained from each pallet.

Where aluminium is used as a mould, in other processes care should be exercised in two respects:

1   The aluminium should be anodised or the mould run with a couple of dummy casts to promote the formation of an oxide protective layer.
2   Reinforcement, whether galvanised or not, should not contact the mould otherwise bimetallic reaction will occur unless something like

potassium chromate at a concentration of 0.001 per cent m/m cement is added to the mix. The exact required amount can be calculated from knowing how much chromium the cement has in it. This, if not present, needs to be increased to about 10ppm.

3.2.6 Composite

Often more than one type of material is needed for a mould, due to the main material not having adequate performance. Ignoring cases where timber moulds are painted, the following are examples of composite applications:

1   a GRP panel mould might well need to have a plywood, blockboard or similar base under the GRP to prevent bowing as well as timber or steel frame edges for the same reason as well as to resist abrasion;
2   a timber mould might need lining with controlled permeability sheets[43] for the aesthetic requirement of few or no surface imperfections;
3   concrete moulds need steel or similar inserts attached to their reinforcement for mechanical handling requirements.

3.3 Treatments

The sub-sections that follow deal with materials but a few preliminary words concerning application might be found useful. Labour-intensive processes often involve the use of mechanical or hand-held compaction hammers or vibrating pokers. These striking the sides of a mould can damage the surface and the use of rubber or similar caps on the tool heads will inhibit the likelihood of damage. It needs to be borne in mind that edges and corners of a product need good compaction and mould contact with the compaction tool is likely. For many compacting tools these caps, or 'policemen', are available as proprietary items.

3.3.1 Release agents

The main purpose of a release agent is to stop/inhibit the concrete from sticking to the mould and is a general requirement except, possibly, for wet-cast concrete in PVC-lined or individual moulds and for moist mix design cast stone in timber moulds. All that is theoretically required to debond the concrete/mould interface is a monomolecular-thick layer but this is not possible due to practical reasons of application as well as mould texture. However, it points to the necessity of being as frugal as possible in application as generosity does not result in improved release and commonly leads to retardation of hydration as well as staining. In order to deal with over-application, it is often of benefit to allow the excess release agent to collect in the corners of the mould and rag off the excess. Alternatively, apply the agent with an oil-wet rag.

At least five problem can occur when a release agent is misused/poorly applied/poorly selected:

1   The concrete will stick to the mould and suffer damage on demoulding.
2   The surface will have a patchy or stained appearance.
3   Over-use can result in chemical retardation of the setting and hardening leading, possibly, to dusting and/or easily damaged surfaces and arrisses.
4   There might be a detrimental reaction with a mould paint.
5   Rusting of steel and swelling of timber may be promoted.

Another aspect related solely to the release agent is that of health and safety. All release agents are chemicals and some personnel may be sensitive or allergic to particular types. It is advisable to wear protective gloves and also of importance to ascertain operatives' personal or family history to all chemicals (including cement) likely to be encountered in production.

There are five basic groups of release agents:

1   non-emulsifiable mineral machine oils;
2   emulsifiable oil-in-water phased oils that can be extended with water;
3   emulsifiable water-in-oil phased oils that are immiscible with water;
4   metallic stearates and similar soaps known as chemical release agents;
5   lanolin creams and waxes.

There are also release agents based on vegetable oils known as 'VERA' which are claimed to be non-toxic, biodegradable, safe to handle, easily sprayed, non-flammable and having high coverage capacities. In the past decade experiments have been carried out in Holland on construction sites and in precast concrete factories using water-based emulsions and neat oils. Good results were said to have been obtained. The claim about the non-flammability of a neat vegetable oil is questionable; it might well have a much higher flash point than conventional mineral oils. No mention was found of mould or other growth potential but the potential for the use of VERA has to be a future option if the availability of petroleum products becomes more restricted.

As emphasised earlier, neither too little nor too much of any of these agents should be used as one or more of the above-mentioned defects can arise. Typical coverage rates range from 15–30 m²/litre with airless spray application or similar coverage with a wet rag being obtained. Brush application followed by a drainage period then removal of excess with a rag is also acceptable.

Release agents of types (1) or (2) can be used if there is little or no regard to the surface appearance of the finished concrete, nor effects on the mould, its paint or anything else. The more expensive (3),(4) and (5) types are cheaper to use than (1) or (2) because they are applied more frugally and lead to less surface making-good operations. The lanolins were used for many spun products such as pipes and lighting columns as they are extremely stable under the spinning forces. They still have an application in many bespoke architectural units such as fireplace units as they produce one of the best surface finishes. Table 3.1 gives advice regarding the use of agents on different types of mould when plain concrete is under consideration. For architectural and light-coloured concrete types (3), (4) or (5) should be used.

3.3.2 Paints

Apart from steel and plastics moulds paints can considerably increase the performance and lifetime of moulds. The first edition of this book went to great lengths describing the results of years of laboratory and precast concrete works research under the auspices of the British Precast Concrete Federation and the results are summarised and updated in this section.

Irrespective of various claims that are made about the vast range of paints available (at the author's last survey there were more paint manufacturers in the UK than any other product one could name), the presence of a pigment in a paint vastly improves its performance. As an example, a pigmented shellac-based paint can give a performance comparable to a polyurethane clear lacquer. The other advantage of using pigmented paints is that in a two-coat system one can use different colours, making it easy to see when the top coat wears away as the first coat colour becomes visible. Three salient rules put the field of mould painting into perspective:

1   The paint system must be compatible with the substrate onto which it is to be applied.
2   The paint should always be pigmented as this contributes more to its attrition-resistance than the type of paint itself.
3   High gloss finishes should be avoided as they promote hydration staining discussed in 3.4.1.

Table 3.2 typifies the number of uses obtainable for the common applications

Table 3.1 Suitable release agents for various types of moulds

Mould	Agents
Unpainted wood	(1), (3), (4)
Painted wood†	(1), (2), (3), (4)
Unpainted steel	(1), (3), (4), (5)
Thermoset plastics	(3), (4)
Thermoplastics	None or (3), (4)
Aluminium	(1), (2), (3), (4)
Concrete (painted)†	(3), (4)

† Some types of paint can be degraded by type (4) agents and trials are necessary when doubt exists.

Table 3.2 Approximate number of uses (to nearest 5) for various treated moulds

Paint	Mould or lining	Unpigmented	Pigmented
Epoxide, polyurethane	Non-resinous wood	45	100+
	Hardboard	20	55
	Resinous pine	3	8
Chlorinated rubber	Non-resinous wood	20	30
	Hardboard	35	50
	Resinous pine	35	60
Cellulose, oil, shellac	Non-resinous wood	15	20
	Hardboard	10	20
	Resinous pine	1	5

of paints to timbers and hardboard. Users need to be aware of health and safety requirements in paint usage and application. This might mean that mould would have to be sent to specialist companies for the work to be undertaken.

Choice of the type of paint largely depends upon how many uses are required and the type of timber or lining material used. Pigmented epoxies or chlorinated rubber-based paints can give up to 100 uses with the chlorinated rubber-based paints having a superior performance when there are migratory resins in the timber.

Consideration can also be usefully given to applying a cheap paint to the non-concrete contacting surfaces of a mould as this would promote less concrete and other material adherence.

Painting steel moulds might only be considered when a mould is to be put into storage for a considerable period. The non-concrete contacting part of the mould can be degreased with a water-soluble detergent, dried, treated with phosphoric acid (all strictly to the supplier's instructions), then coated with a single coat of a pigmented epoxy paint. The concreting part of the mould can be treated with a mould release agent and, for extra protection, rendered with a sand/cement mortar and left.

3.4 Problems

3.4.1 Hydration staining

The incidence of this virtually irremoveable and most unsightly of aesthetic defects continues to be experienced. It has been observed on both in situ as well as precast concrete although the complaints recorded have referred to precast concrete products mainly. The probable reason for this is that the surface finish requirements and expectations for precast products are more stringent than those for in situ concrete. In addition to this comparison, there is another exacerbating factor with precast products and that is that they are commonly stacked at an early age. Since both mould smoothness or polish as well as the effect of stacker packs have been found to be significant factors in causing this problem, it was considered advisable to give the problem a high degree of prominence.

Several factors associated with hydration staining are listed but, before discussion, the underlying mechanism previously explained[44] as the cause of this problem probably needs to be re-stated.

The words 'macro' and 'micro' have been used at various places in this book and will be found in many other treatises dealing with concrete and other materials. Descending about four orders of magnitude below the micron, atomic forces are encountered at the Angstrom level of length measurement. Strong levels of atomic attraction, due to Van der Waal forces, are known to operate at these atomic distances. A relatively smooth surface of a mould or a stacker piece is likely to have areas of relatively atomically-smooth surface. Therefore, fresh or fairly young concrete in contact with such a surface will tend to adopt the same macro and micro (atomic) smoothness as the surface with which it is in contact. A relevant and possibly well-known manifestation of the strength of these forces may have been experienced in the mounting of 35mm slides. In addition to the 2No. plastics locking frames for each slide, there is provided a stack of tissue-wrapped smooth glass plates of which two are used to encase each transparency. It will be noticed that the removal of each piece of glass can generally only be achieved by a sideways sliding action. It will be found almost impossible to pull a piece of glass off vertically along the same axis as the stack of glass plates.

Another example of these forces but one not likely to be generally known is when two pieces of pure copper have their ends polished to a mirror-smoothness in a nitrogen box (to avoid oxydation) and these polished faces are brought together under light hand pressure. A bond as strong as a brazed or soldered joint is obtained and the two pieces of copper cannot be parted.

The following points are the main factors involved:

- concrete shrinks during the hydration process of the cement;
- the shrinkage away from the mould or formwork leaves a small gap large enough for air with its small carbon dioxide content to ingress;
- for typical OPC concrete this results in a light grey colour due to slight surface carbonation;
- concrete cured under air-free conditions exhibits a colour between dark grey and black;
- white Portland cement concrete and cast stone exhibit a light blue/grey colour;
- smooth polish-finish, gloss-painted moulds or formwork will promote hydration staining as atomic smoothnesses can be obtained over relatively large areas;
- smooth-faced packing pieces or stacking blocks placed against visual

faces before the concrete is about 2 days old will promote the same problem;

- hydration staining has been found to have no relation to the type or amount of mould release agent or oil used;
- the glossier the finish of the mould or formwork, the worse the problem;
- the strong attraction forces between the concrete and the mould or formwork make demoulding or stripping difficult;
- weathering does not ameliorate the effect; if anything, the concrete surrounding the stained area tends to lighten in colour while the stain remains virtually unaffected. This makes the stain stand out more in contrast.

Hydration staining is identified as dark glossy area(s) on the surface accompanied, commonly, by difficulty in demoulding/stripping. This staining typically penetrates 10–30mm deep into the concrete. Weathering, if anything, tends to emphasise the contrast between stained and unstained areas. Hydration staining has been found to have no similarity to the staining caused by leaves, planks of timber used for stacking after the concrete is a few days old where saponification by the lime in cement generally causes such stains to fade or disappear.

Once hydration staining has occurred, there is no direct remedial action that can be taken. The area(s) affected can be cut out and made good but the aesthetics of the made good concrete would need to be discussed. The alternative indirect treatment would be to paint the concrete with a low maintenance paint such as a silicate or acrylic-based system ensuring that the surface can still breathe, i.e. allow the passage of water in vapour form. This treatment would give an overall darkening effect helping to mask the areas of hydration staining.

Do not use high gloss finish moulds or formwork. Where paints are used, they should be matt finish and pigmented with the pigment colour changed if more than one coat is to be applied. New steel moulds often benefit by having a couple of dummy casts to roughen the surface. There is no need to fill the mould completely, a thin cementitious rendering would suffice.

Where the mould or formwork has a high gloss finish, two or three consecutive daily mortar applications and removals help to 'weather' the surface into a consistent use behaviour. The alternative is to lightly sandblast the surface so as to produce a matt finish.

Where stacking pieces are to be used, they should not be deployed before the concrete is 48 hours old. Even after this time, for visual faces, the stacking items used should preferably allow part-air ingress. For example, expanded polystyrene blocks would be preferable to polythene-wrapped pieces of timber.

The associated experience of the concrete sticking to the mould or formwork can often be alleviated by the incorporation of a valve into which compressed air is blown during demoulding and/or stripping. The sticking would be due to the Van der Waal attraction forces which, at a simulated total vacuum suction level could reach 0.1MPa. Over an area of concrete of, say, 0.1m², the force required to release this area of adhesion would be about 1T.

A short experiment to exemplify hydration staining is to fill two jam jars or similar transparent glass containers with concrete or mortar, compact it and seal one of the jars with an airtight lid. The comparison will be observed the following day when the mixes have hardened.

3.4.2 Release agent staining

This was mentioned in Section 3.3.1 and is usually observed as dark-coloured patches on the surface of the concrete, especially predominant at arrisses where the release agent can tend to collect. Most type (1), (2) and (3) release agents are saponifiable (a soap-forming reaction with the cement) and tend to bleach out with weathering.

Attempts to remove the staining by chemical means should have their efficacy tested experimentally as many cleaning agents can make the appearance worse. The use of hydrochloric acid should be avoided if at all possible. If acid has to be used, however, three steps are recommended:

1   Thoroughly wet the surface first with plenty of water for at least ten minutes.
2   Apply as weak a concentration of acid as possible to achieve the desired effect.
3   Repeat Step 1.

The reasons for this is, first of all, to inhibit dry or nominally dry concrete sucking the acid in depth by capillary reaction so that the etching is promoted in the surface region. Second, by inhibiting the presence of acid from the region of the reinforcement, corrosion is prevented. Third, by removing all residual unused acid by the repeat washing, the surface becomes less hydrophilic when dry, thus lessening the possibility of lime bloom.

3.4.3 Release agent retardation

Many release agents act as chemical retarders to the setting and hardening of cementitious mixes and excess application, as mentioned earlier for all release agents, should be avoided. Remedial treatment is extremely difficult as defective material has to be removed and the resulting void made good. This will generally show up as a patch repair.

3.4.4 Bimetallic corrosion

Readers might be familiar with the periodic table in which all known elements sit but all metals sit in another table known as the electrochemical series. This series is not in the same order as where the metals sit in the periodic table. In simple terms what this means is that a metal in a conducting liquid such as water will set up a potential voltage difference if that metal is touching or in close proximity to another metal with a different electrochemical potential. The resulting voltage difference will cause a flow of metallic ions from the metal with the lower potential either into solution or towards the other metal.

For all moulds, ungalvanised steel tie wires wrapped round galvanised reinforcement while the concrete is moist will cause corrosion in the steel to occur over the small areas of wire contact. The main problem area is the case of steel moulds where galvanised reinforcement acting as kickers or continuous bonding for in situ joints travels through holes in the mould end plates and is therefore in electrical contact with the mould. An electrical field is set up that manifests itself as soon as concrete is introduced into the mould. Hydrogen gas evolution is one of the elements produced and this occurs over the zinc surface of the coated steel. This results in a detrimental effect on bond to the concrete.

As suggested before, one way of inhibiting this reaction is by a chromate addition to the mix such that what is added together with the contribution from the cementitious materials gives a total chromium ion content of the order of 10ppm m/m cementious content. An alternative approach is to use rubber or other insulating grommets between the reinforcement and the mould sides.

3.4.5 Double clamp-on vibration

Some precast units have their concrete compacted by more than one clamp-on vibrator and this has been observed for column/half-roof rafter units. Most vibrators are powered off of 3-phase electricity and, where two or more vibrators are used on a mould, it is essential that both vibrators are wired into the same phase.

At the precast works in question this error resulted in a robust steel mould splitting. It is doubtful if a timber or composite mould could tolerate this treatment and trials are not recommended.

3.4.6 Mould storage

This matter possibly does not receive as much attention as it should bearing in mind that moulds, especially those with complex shapes and rather bespoke in nature can cost many thousands of pounds each. Storage buildings obviously need to be secure but it is also a good idea to have some form of temperature control to keep mould temperatures in a range that does not give rise to distortion in the form of warping or bowing or similar. Where timber moulds are stored, humidifiers are also advisable. Timber has an equilibrium moisture content in the range 15–20 per cent m/m. Timbers will warp if they dry out or become too damp, coupled, in the latter case, with the risk of wet rot.

Timber moulds of softwood should be painted (see Section 3.3.2) on all sides as this not only assists in keeping the moisture content uniform but protects them from detritus. Steel mould should be treated with a chemical release agent on all sides and, provided that the floor load capability and the requisite handling plant is available, filled with concrete.

Thermoset plastics moulds generally need no precautions apart from environmental ones mentioned above. Thermoplastics moulds and linings will suffer in the long term from plasticiser migration and are generally not suitable for long-term storage. It is probably better to consider a recycling option as PVC, as an example, can be melted, plasticiser added and reformed.

4 Production control

4.1 General

There are several reasons why, in this second edition, a different approach to the discussion of production control has been adopted and these reasons have been listed under the five main headings that follow. For further and in-depth discussion, the reader is referred to Richardson's (1991) publication.[45]

1 People do not have as much time to read technical books in depth as they used to. In addition to this, many individuals seem to exist in isolated worlds and publish or write or say things that have been published before. People appreciate guidance and ideas set out in as succinct a form as possible. The reason for this is that the number of technical books available on any given subject is often large whereas about twenty years or more ago the choice was rather more limited.

2 It is common in teaching, training, lectures, etc. to use bar charts, bubble or key markers, with key words often selected to produce gimmicky acronyms. What has to be emphasised and borne in mind is that the reader has to be kept aware of the target and continually directed to it.

3 Since the 1982 publication of the first edition, large sections of the industry have ceased to exist. Well-known household names have disappeared and several types of production no longer are undertaken. In addition, there have been many takeovers. Although this has possibly led to a more compact industry, it cannot be stated that the smaller back garden shed manufacturer has also disappeared. In effect, purchasers

can still go for the cheapest tender if no thought is applied to cost effectiveness.

4   The use of personal computers has grown beyond all recognition and these can and are being used for a large range of activities such as:

   (a) computer-aided design in drawing office work;
   (b) costing of tenders with inflation and material/labour adjustments built in;
   (c) sales and storage and delivery controls;
   (d) materials and mix controls coupled with material re-ordering processes;
   (e) financial control including salaries and wages coupled with weekly/monthly printouts and bank transfers;
   (f) staff training, re-training and auditing procedures;
   (g) plant maintenance schedules;
   (h) test and quality control data;
   (i) document, including procedures, control;
   (j) complaints/actions and preventive activities.

These ten items are not meant to be comprehensive but illustrate some of the many activities that PCs can undertake. Tied up with these many advantages is a major disadvantage and that is complacency through relying upon a piece of software that appears to be doing its job. What is so often lacking is 'benchmarking'. The software used to drive the PC is human in origin and subject to the errors that a human might make. Therefore, any program, purchased or 'home' produced, should be put through all its planned actions in dummy runs before it is accepted into the system. For example, just because a mixer console states that a skip of concrete has gone to shed number 3 doesn't necessarily mean that it has. Just because a re-training schedule is called up for operative 'x' might have missed out on the fact that 'x' left the company 6 months ago.

   Having emphasised the need and value of benchmarking, each benchmark undertaken needs to be documented into a benchmarking file. The fact that a particular aspect has been benchmarked does not mean that it is faultless for all time. Re-benchmarking is advisable when program amendments are made just to check and be sure that one's aims are being achieved. The joint keywords of any activity are 'documentation' and 'traceability' and these words are neither gimmicks nor stand-alones. In addition, it is rather pointless having documentation without effective traceability as well as the opposite. All documents held in electronic or hard copy form require a file reference and a date on them with a further reference as to the document superseding a previous one if relevant.

4.2 Staff

Although each member of staff has terms of reference for their tasks, each is a member of a team whose target is to manufacture and deliver precast products that are of an agreed quality and price and delivered at the agreed time and place. Therefore, without knowing in-depth details of all the activities in a precast concrete works, it is reasonable to expect each member of the team (all members of the works, salaried and waged) to know the basic capabilities of all other members of the team. A few examples of these cross-over activities may be quoted:

1   The purchasing department may well be satisfied with the routine deliveries of materials but may not have studied the acceptability of alternative sources should there be an interruption/unacceptable change in quality/price of the main supplier. The department would need to liaise closely with the technical department to ensure that the a virtually-identical (appearance also being a property) product is on offer.

2   A member of the sales staff may have committed the company to a contract without first checking capability of obtaining materials, inadequate time, lack of storage space, and so on. All too often, cases are heard where lead-in times are not included in a schedule of operations in contract time planning. One recent example of material shortage was that of reinforcing steel due to extra requirements from Far East countries.

3   All too often there is no 'Plan B' should a member of staff leave, go off sick, in order to dovetail staff during holiday absences. In staff's training/capabilities listings it can be of benefit to have each member of staff trained to do at least two tasks. Possibly the most important of these is the person(s) in charge of the mixer on whom so much of concrete quality rests.

4   Rewarding staff for useful suggestions . . . plus many more.

Before discussing training in more detail, this, together with all the other activities in the construction industry, relies upon enough work coming in to keep a business viable. In this sense the industry has probably been too reactive to events and not pro-active enough. The industry's work input relies to a large extent upon government and, indirectly from government via Local Authorities. This is coupled with input from industry and the private sector. The crux of the problem is that governments, irrespective of political persuasion, have used the construction industry as an easy and quick-to-respond financial regulator for decades. This leads to the question as why one should train people when one might have to cut back on work at any time and, generally, with short notice. Thus, it behoves the construction industry to adopt a much fiercer approach and be pro-active so that we all have a sensible degree of control over our future.

Training forms a most important part of the activities of any member of staff from managerial through to operatives. Procedures for training should form part of the company's Quality Manual and both their initiation and re-training needs to be scheduled and recorded. This is one of the many activities carried out jointly between the personnel department (currently called 'human resources') and the person in question's line manager. The Managing Director or Chief Executive would often have no line manager but this does not preclude him or her from self-discipline in regards of training by attending up-dating courses, trade federation activities and networking with colleagues and competitors.

4.3 Plant

Anything from a hand trowel up to a multi-million pound machine as well as control consoles, switches and the like are all plant and require control in one form or another. There are two main aspects of this control:

1 documented schedule of all requirements with a call-up procedure for both in-house (e.g. maintenance) and out-house (e.g. manufacturer maintenance, calibration) activities;
2 regular non-production downtime for cleaning, oiling and other actions aimed at having everything as ready as possible for a trouble-free start to the next production run. It is suggested that these activities be treated as part of the work regime with some form of possible added incentive.

Illustrations can be shown of some of these aspects and both good and bad practice ones have been selected as it can be argued that more can be gleaned from viewing the bad rather than the good.

Figure 4.1 shows a mix control console, however, it has been placed next door to manufacturing plant and thus subject to air-borne dust and other detritus. These consoles are not cheap and are not only best placed in a separate and enclosed room but a polythene or similar covering over the controls would inhibit ingress of dirt/dust and, at the same time, avoid dirtying from hand contact.

The mixer shown in Figure 4.2 would also have benefited by having been placed in a separate enclosure as, even with the best will in the world, they do produce dust. The mixer shown in Figure 4.3 is a small pan mixer used outdoors but with having been placed under a tree, was subject to the addition of unplanned ingredients, especially during autumn.

On the other hand, the delivery skip shown in Figure 4.4 shows a well-maintained piece of plant where all the controls are clean and visible. It is too often the case for plant that when maintenance is required the details of manufacturer, serial number, etc. cannot be read because of a long-term build-up of concrete.

Health and Safety, Risk Assessment and sundry other effective and

Figure 4.1 Poorly-sited mix control console.

Figure 4.2 Pan mixer placed in open works area.

Figure 4.3 Pan mixer placed outdoors under a tree.

Figure 4.4 Modern skip delivery of concrete to a casting shed.

buzzwords play a key role in the use of plant as with all other works activities. Strict and documented procedures should be laid down and understood by all operatives. Failure to observe any of these rules is a disciplinary matter and should be recorded. If, for example, ear protectors are provided for work in high noise level areas and an operative refuses to wear these, this refusal should be recorded.

4.4 Materials

This particular sub-section is not overlong as there are numerous publications on the subject. As far as precast production is concerned, most of this has been discussed in Chapter 1 and there is no need for repetition. Materials are seldom to blame for things that go wrong; it is the way they are used or mis-used. When the author was employed by Laing R&D, all trouble-shooting enquiries and activities were grouped into one of three categories:

1 Material
2 Design
3 Workmanship.

Even though the division between the last two was not always that distinct, all enquiries were listed over a ten-year period and materials were found to constitute 14 per cent of the total. As an example of categorisation, a material specified not to be used below a temperature of 10°C and failing at 5°C is a design failure, not a materials one.

Aggregates, cement, additives, water and admixtures are the main ingredients in concrete and these may either be routine repetitive deliveries or 'specials'. Special materials often get special attention as they are generally used for architectural products. This special attention may take the form of a thorough investigation of properties, a visit to the source, a view of other buildings where it had been deployed, and so on. It is not suggested that this be done on a daily, weekly or monthly basis for routine deliveries but that complacency be avoided where it could be dangerously assumed that there have been no significant changes delivery to delivery.

Both routine and special materials found to be satisfactory should be representatively sampled and stored in sealed and dated containers. Thus, if there is any query about changes in supply, one has at least one reference point. Furthermore, each routine delivery of aggregate should be checked before permission be given to unload. Cement and additive deliveries cannot be so checked but the delivery note needs to be carefully examined to record that what has been ordered is that claimed to be in the delivery vehicle.

Reinforcement should bear the CARES bar marking and obtained from an approved source. Prestressing wire/strand spools should each be labelled with a BS.EN Certificate.[46]

Any source of material should be from an approved purchase list examined (and re-examined regularly) and found to be acceptable. It should not be assumed that because a supplier has a BS.EN.ISO.9001 Certificate that what is on offer is of a suitable quality. That quality standard is a management document and is generally only reflected in the product by its consistency of property or performance.

Precasters who manufacture their own timber moulds are probably in not so fortunate a situation as they have little control of timber quality, including, most importantly, its maturity. It can only be recommended that suitable sources be identified, used and continuously checked. The moulds made should also be routinely checked for dimensional tolerance and stability and taken out of production and re-fettled or disposed of if necessary.

Admixtures, additives and hardware need to be treated with as strict a control as possible. The first two are covered by ENs and conformity certificates would need to refer to these. Products such as cast-in sockets, spacers, fittings and fixings are not necessarily covered and such materials found to be suitable should be sampled, documented and kept in sealed containers for future comparison.

4.5 Product

The final manisfestation of all the foregoing activities is piece of hardened concrete weighing anything from a kilogram or two up to many tonnes. This product has to go through several operations from the mould to site before it can be labelled as fit for purpose. Comments on these operations can be given as recommendations in a staged format. One of the most important matters to be settled at the preparation stage is which way up is the product best manufactured – face down, face up or on edge? The decision is largely a function of geometry and surface finish but will also dictate how the reinforcement is to be made, supported on spacers or suspended from a head frame or other design. Coupled with this will be decisions on lifting, handling, stacking and transport to the site. The following are the key points in production control.

Materials need to be stored in sensible fashion with computer-controlled mixers being re-programmed when there is a storage change. Manual or automatic weigh and volume batching equipment needs to be on a strict maintenance and calibration schedule. Some advanced automated mixer processes not only control the required mix but also, when bins or silos get below a certain level, will re-order more material. All such processes need to be checked and benchmarked on a regular basis.

The mix water content is probably the most important item in material control and this can be controlled in several ways. Water can be added according to what remote in-aggregate moisture meters record. The W/C(T) is computer achieved by metering extra water over what is already in the aggregates.

4.5.1 Demoulding

Depending upon the process, the product may be anything from seconds to hours old at the time of demoulding. Freshly demoulded products such as hydraulically pressed paving/kerb, packerhead pipes, blocks, floor tiles and earth-moist cast stone need special attention as they can be easily damaged. The most critical of these is cast stone as it is the only one that is manual and not machine-intensive and thus not protected to a large extent by the plant and/or the process. The only pro factor in the case of damage to cast stone is that the product can generally be made good while fresh but this requires a great deal of expertise.

The problems in demoulding hardened concrete generally relate to arris damage due to the arrisses being weak compared to the bulk of the product with much of this being due to either poor compaction and/or inadequate curing. Making good arris damage can be undertaken using a low polymer content mortar but this might well be visible and unsightly, especially in the case of a visual concrete (architectural) product. Chapter 11 deals with this subject in greater detail.

Sticking to the mould and hydration staining was discussed in Section 3.4.1 and can be a problem in demoulding. The problem is largely associated with high gloss moulds and is likely to occur irrspective of what and how much release agent has been used. The recommendation is repeated that high gloss finish moulds should be avoided. Large plain areas of glossy moulds can also cause problems due to interface stiction and some of this can be alleviated by having compressed air valves built into the mould preferably in regions of the plain face where the valve marks would not be too much of a visual concern.

4.5.2 Stacking

It is assumed that operatives involved in stacking/storage are trained and familiar with all safety aspects and, where applicable, familiar with and able to use crane signals. A hazard can exist in the use of woven wire loops whose internal wires are difficult or impossible to inspect. Although the outer wires may give the appearance of pristine newness, the inner strands can be corroded. It is probably a good idea to store these in an oil bath when not in use. An indication of possible trouble can be obtained by placing the loop in water for a few minutes then leaving it to drain on the floor. Rust diffusion should be easily spotted and at any sign of trouble the lifting loop should be scrapped.

Where stacking is in the form of product on top of product, stacking blocks need to be in the same vertical line and at fifth points for uniform reinforced units and near the ends for prestressed products. For visual concrete (which includes cast stone) stacking blocks need to allow 'breathing' over most of the visual face contact area otherwise permanent hydration staining will result (see also, Section 3.4.1).[47] It is also important in stacking a large number of repetitive products to have their manufacturing dates identified so that the oldest are delivered first.

If units such as paving flags have to be stacked on edge, as distinct from being banded on a pallet, they should rest against a rigid wall or doubled opposite each other with stacking and offloading in reasonably equal amounts from either end. For visual units stacked in contact on their edges, it can be of benefit to have small stacker pieces at their heads to inhibit concrete-to-concrete contact.

4.5.3 Delivery

In addition to the comments made in Section 4.1 it is important to bear in mind that what arrives on site should be in the same condition as when it left the works. Allowances need to be made for the route to be travelled, vibration effects and details of offloading, and so on. Even if it means that for the sake of achieving perfection, underloading transport is preferable, then so be it. Figure 4.5 illustrates an underloaded trailer delivering cast stone units to a site. The units are well stacked, have spacers between them and are secured on pallets to keep them clear of the trailer bed with added vibro-insulation.

Bespoke A-frames are popular for transporting cladding panels as they are in their optimum structural position and can be rested on insulating pads with interspaced spacers between each to prevent concrete-to-concrete contact. As mentioned earlier, if cladding units contain window openings, the

Figure 4.5 Lightly-loaded trailer with well-supported cast stone units.

units are likely to attract more handling respect all round if the glass is fixed in the works and remains visible at all times.

4.5.4 Installation

Products installed on site are all too often treated with a large degree of disdain. There are two main things the producer can do about this:

1 Submit tender with 'recommendations for use' on site document.
2 Ensure, and if necessary formally warn, that products are delivered as mature as possible and, if demanded too 'green', advise of things such as continual hydration with its accompanying shrinkage and weakness of arrisses irrespective of the strength of the product.

It is important to bear in mind that because a cube strength result or other strength indicator shows that a product is strong enough to be transported, all properties will have become static. Although the strength will continue to increase due to increasing hydration, shrinkage will continue to occur and permeability will continue to improve. The shrinkage effect can manifest itself with cracking problems if a unit is rigidly fixed at its ends as well as mortar joints opening up in masonry work. The permeability improving effect could well mean that products delivered at a young age would be more detrimentally affected by staining agencies than those delivered when more mature.

4.6 Client

One might well ask what the client has to do with production control and, as one thinks more about the subject, the answer is 'a lot'. In addition to discussing the points of relevance it has to be admitted that there is a dearth in technical books on the client when he or she is the person who gives or can give the order for work. With transparency of all operations relevant to control being the target, a client visiting a works witnessing a slick production control in place:

1 is highly likely to be confident to placing an order;
2 is equally confident during a contract that the quality required and the schedule agreed is reached and maintained throughout.

Furthermore, clients, especially architects, can put their own ideas forward regarding modification to processes, production of samples for new forms of precast product, and so on. For some types of production such as ashlar and cast stone, he/she might quite easily request less control in one of the processes (viz. mix design) in order to simulate the natural stone-to-stone variations that one gets in many forms of natural stone. He/she could also opine that a partly mixed addition to the completed mix of a black or off-black mortar could well read out in the finished product as one of the natural bed veins common in volcanic-source rocks.

Although complaints strictly come under one of the ISO 9001 items, that Standard only lays down the documentation and action aspects and does not really cover human aspects. A golden rule in dealing with complaints is to let the client spell it all out without interruption or arguments or any committed form of facial expression. The only time one should speak is when qualification or details of the complaint have not been conveyed. If the complaint is justified, the first thing to do is to apologise without qualification and to offer appropriate restitution. If the complaint is only partly justified or unjustified, full reasons should be given and, at the same time, apologise for any misunderstanding. All too often relationships can be jeopardised by mishandling of clients and tact and diplomacy should be used at all stages.

5 Labour-intensive processes

5.1 General

Unlike the first edition, this chapter does not list typical mix designs as this was not considered to be particularly helpful. Specifiers either tend to use a strength classification or rely upon their chosen precaster producing a regular supply of an acceptable and repetitive product.

For run-of-the-mill products the precaster will generally use locally supplied incoming materials because of their cost-effectiveness. For bespoke products such as visual concrete and cast stone, aggregates and cements are often transported over relatively large distances and sometimes even have to be imported.

This chapter groups together all forms of precast concrete manufacture where there is no machine in use that is specific to the product. Thus, no matter what mixer, compaction and finishing tools are used, the labour content is intense and constitutes a significant part of the cost in manufacture. There are some grey areas between labour-intensive and machine-intensive processes and these are discussed in the next chapter. The descriptive dividing line adopted between the two process groups is that where operatives generally only become involved at the stacking and/or loading stage, then that is machine-intensive.

Irrespective of the attraction of installing fully-automated production processes wherever possible, there is a limit to what can be installed. The things that cannot be overcome by automation are generally size and/or shape and/or intricacy and/or bespokedness. The precast concrete industry has to be in the position of being able to offer the client anything that can be made in precast concrete varying from a 1kg garden gnome up to a 1000T bridge caisson. Only labour-intensive processes and products are considered in the rest of this chapter.

Here far greater prominence is given to cast stone manufacture than was given in the first edition. The demise of household names such as Empire Stone, Minsterstone, Girlingstone, Bradford's and Stent over the past few decades has left a gap that has yet to be filled. The industry has been making cast stone since the early twentieth century and there are many edifices and examples still performing well. The industry also produces hybrid products that do not go into the 'works' as described by the Construction Products Directive. Thus, control documents such as Standards and Codes need to keep this in mind in their formats. It is to be hoped that the contribution of this particular volume will result in a far more adventurous approach by both architects and manufacturers.

5.2 Wet-cast

5.2.1 Prestressed beams and columns

Most of these products tend to be pretensioned and made by the long-established long-line process where moulds are in a long line with the same wires or strands continuous through their ends and anchored at one end and prestressed at the other. Bed lengths can be typically 100 metres long and contain several lines, each of which will contain several moulds, with the number of moulds depending upon the length of each mould.

Concrete, generally from a central mixing area, has to be transported to the casting area, placed into the vibro-insulated mould and compacted by mould-fixing clamp-on vibrators. When each unit is compacted, the vibrators are moved onto the next mould and the process repeated. Once the concrete has reached the required strength, ostensibly determined by cubes cured under the same conditions, the prestress at the stressing end is released, the wires and/or strands are cut or burnt through and the units taken to their stacking location.

Against the labour-intensiveness of the process can be listed its versatiliyy and its ability to tolerate stoppages. Examples of the versatility can be seen in Figure 5.1 showing a couple of I-beams with top bonding reinforcement and Figure 5.2 showing very large bridge beams with both pretensioning as well as post-tensioning near the top of the section.

General aspects of health and safety were listed in the previous chapter and prestressed concrete manufacture has to be treated with extra respect as there is a lot of energy in a prestressed strand or wire whose unplanned and sudden release can lead to fatalities and injuries in both the precast and in situ industries. Personnel should never stand in line with the wire/strand, especially during the prestressing operation. Beds should always have holding-down bands across their tops and end-mould guide plates should be edge-bevelled to avoid nicking the wire/strand. These are some of the main precautions.

Figure 5.1 Wet-cast process I-beams.

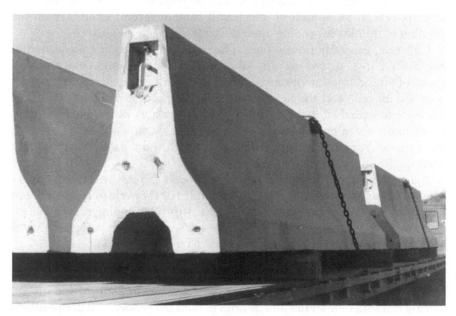

Figure 5.2 Wet-cast process prestressed beams.

Post-tensioned anchor points concentrate a lot of punching shear stress in a very small locality and this is another reason for extra reinforcement in some cases. This reinforcement could be a spiral of high tensile steel bar surrounding the strand or wires. The need may also arise for pretensioned concrete anchor zones but the punching force is over the transmission length and builds up from zero where the wire or strand enters the concrete up to the position where full prestress transfer is obtained.

5.2.2 General reinforced and unreinforced units

This group covers a wide spectrum of products such as cladding panels, ducts and duct covers, tanks, room units, bespoke columns and beams, posts, stairs (singles or flights), and so on. None of these products lends itself to a machine-intensive approach because each is unique and bespoke in its own right. The range listed is not exhaustive and exemplifies the versatility of the precast concrete industry in its ability to meet a vast range of needs. With a degree of crystal ball gazing coupled with imagination there could well be other products either briefly tried or never before made that could be made in precast concrete. One in the former category that comes to mind tried several years ago but, apparently, never pursued commercially, are machine tools. These were thought to have been developed in the then USSR and used for pressing metal plates in car plants.

Examples of currently manufactured products are shown in Figure 5.3

Figure 5.3 Poor lifting on a manhole cover.

which illustrates a 2m diameter (not the access hole) manhole cover where causing minimal damage during lifting was not something that was predominantly in mind. Figure 5.4 shows a large domestic cess tank being demoulded. The last illustration also shows that the product was manufactured in the upside down position to its intended use. This is an important matter and possibly the first one to consider when manufacturing wet-cast products.

This last comment leads to three important questions needing to be addressed at the planning stage:

1   Which way up is the unit to be cast?
2   What are the important parts of a product relating to both dimensions and appearance?
3   How is the product to be demoulded, lifted and stacked?

If the cess tank in Figure 5.4 was to be made in its as-used position, filling with concrete would be a problem as not only would there be difficulty in achieving compaction in the base but many manufacturers would tend to introduce the concrete at more than one edge. At the approach interfaces of the advancing concrete will be an amount of material (viz. dust/release agent) that will be trapped and show as an immovable stain on the finished product. This might be of little concern in the case of a cess tank but if the product was a coffered ceiling visual concrete unit, then it would be a different matter.

Figure 5.4 Demoulding a cess tank.

Section 5.3 deals with earth-moist-cast products of which cast stone is the most noteworthy example. However, due to their often-encountered intricacy, as in the case of a garden gnome, wet-cast in a vinyl or similar pliable mould, could be used. Architectural attempts to produce a natural stone appearance on a large structural product have always been open to argument.

Where disputes do occur is generally for wet-cast architectural cladding where the following problems will always obtain:

1   There will never be a 100 per cent sameness product-to-product as no manufacturer can control aggregate and cement properties and water content as well as curing to such critical limits.
2   Even kept inside the factory and under cover for a long time, the relative humidity and temperature can only be controlled within a practical but not close range.
3   Unless fixed onto the building direct from the delivering transport the producer has no control on how the units are stored on site.
4   No-one can control the weather conditions.

The US Prestressed Concrete Institute has a large and very active Architectural Precast Concrete Division and has published a full guidance document[48] on the above and many more details on this subject. The document is recommended as essential reading for cladding manufacturers.

5.3 Earth-moist-cast

The whole of this sub-section has been devoted to cast stone manufacture. This oft-neglected subject received what was then a reasonable coverage in the first edition of this book but the intervening years were witness to a slump in the industry. During this intervening period at least three of the largest household names in cast stone manufacture ceased production. The past decade has brought about a resurgence in production.

Before going into detail of cast stone manufacture, there is one matter that needs to be resolved and that is what Coade Stone is. This was described at a Concrete Society Meeting as being cast stone; it is not. Coade Stone is a fired product and therefore comes under the definition of a ceramic. Eleanor Coade was the founder of this enterprise which manufactured its products at a works in Belvedere Road in London between Waterloo Station and the river. Its origins came to light in the excavation work for the Festival of Britain site in 1949–50 when furnaces and grinding wheels were unearthed.

The products were and are very durable. Two outstanding examples of these are the lions which used to stand over the entrance and on the riverside roof of the Lion Brewery that also used to be in Belvedere Road. They are currently at the entrances to Westminster Bridge resting on the West and East parapets. One of the lions suffered some weathering at one of its parts and this was successfully repaired with a Portland finish cast stone repair mortar in about 1980.

Another point worthy of mention is description. Cast stone is defined as a product made from aggregate(s) and cement to be used in a manner similar to and/or for the same purpose as natural stone, Decriptions such as 'artificial stone', 'art stone', 'reconstructed stone', 'reconstituted stone' and 'synthetic stone' still persist but are deprecated. There is not only no point in having more than one description for the same thing but there is also a possible legal pitfall in using terms such as 'reconstructed' or 'reconstituted'. Both these words imply that the name that follows the adjective is the sole aggregate. Therefore, the use of another aggregate or a secondary aggregate with the named finish is really a contradiction of terminology. To use the all-embracing definition of, say, 'cast Portland Stone' or 'Cast stone – Portland finish' or similar allows full licence by the manufacturer to use what he/she likes provided that it is acceptable to the client.

The word 'earth-moist' has been used both here and in the first edition to describe the consistency of the mortar-like mix that is used in this production process. The word 'semi-dry' is still used to some extent but is not scientific-ally correct as the word 'dry' means arid and completely destitute of mois-ture and the prefix 'semi' means half. Since cement needs water to become hydrated, there is no way of making a cementitious product with half nothing for a water/cement ratio.

The 'earth-moist' description in cast stone manufacture means that the person in charge of the mixer has a critical role to play, much more so than the person in charge of a manually controlled wet-cast mix where a larger degree of water content variation is acceptable. In earth-moist cast stone mixes the test is to pick up some of the mix and squeeze it in one's hand when a cohesive 'sausage' should be obtained. This should not break up when hand pressure is removed nor leave more than a dusty mark on one's hand afterwards when the sample is released.

This judgement by the mixer operator is all-important as the aim of the earth-moist product is a mix that, when compacted, produces a unit that can be instantly demoulded, and:

1   has no wet cement laitance on the surface needing extra finishing work;
2   has no additional weakness resulting in friable arrisses and/or poor compaction and inter-particle bonding;
3   resembles natural stone.

There is no magic formula for the water requirement. Mixes used in cast stone production are generally mortars and contain fine crushed rock material such as limestone, sandstone or granite. Very often a natural rounded sand is used with the crushed rock to promote workability. The crushed rock grad-ing not only does not follow the recommended grading of a crushed material one would use for concrete production but should not do so for aesthetic reasons. To produce the fine-grained surface appearance of natural stone, a significant percentage of fines is necessary but not too much as that would cause a high water demand with subsequent weakness. A practical rule of thumb for crushed rock material is that the grading should be approximately equal amounts in each sieve range. However, in-house experimentation is always recommended to ascertain the optimum range for the aggregate in use.

The addition of the natural sand has its advantages and its disadvantages. Natural sand has more movement sensitivity to moisture than crushed natural rocks and the problems of surface crazing and drying shrinkage cracking were discussed at length in the first edition. If there is too little sand, stress centres at the surface can be set up, resulting in surface crazing. If the sand is too fine, then shrinkage cracking is promoted. The two general rules expressed in the first edition are repeated here:

1   The natural sand should constitute at least 25 per cent of the total aggregate.
2   Not more than 5 per cent of the natural sand should pass the 150μm sieve.

Crazing is a surface map cracking pattern that typically has a depth of about 1mm, a crack surface width of 0.2mm and a 'square' dimension in the range 20–40mm. It is generally more of a drawback than for ordinary concrete because not only is cast stone a visual concrete product but it is commonly of a light colour as in the common case of a Portland finish. The apparent surface crack width is generally larger than 0.2mm because dirt-borne water is drawn towards the craze line by capillary attraction, leaving the dirt on the surface.

Drying shrinkage cracks are generally a series of parallel fine cracks at right angles to the long axis of the unit and at a typical spacing range of 100–300mm. The crack surface width varies from about 0.1–0.5mm and the depth from 1–100mm. However, the surface crack width is no indication of the crack width at depth and in both cast stone and ordinary concrete drying shrinkage cracks taper with increasing depth into the product.

Neither crazing nor drying shrinkage cracks are freeze/thaw risks nor have ever been proven to be so. A simple physics approach supports this contention because if a craze or drying shrinkage crack became filled with water and then froze, there is a free surface in which the ice can expand without causing stress. The same mechanism applies to a domestic ice cube box which never breaks as ice forms.

To give the reader an insight into what has been done with cast stone and, at the same time, giving ideas for the future a number of illustrations follow which, it is hoped, will encourage a more adventurous spirit by all the parties involved. In the 1920s and 1930s cast stone was somewhat frowned upon and regarded as a cheap substitute for natural stone. This is exemplified by virtually every building in Regent Street in London where the main façades are in natural Portland Stone and the side and rear elevations are in cast stone Portland finish.

Apart from the staff, pot and its stand, Figure 5.5 shows a statuary group of figures manufactured c.1934 and last known to be in the Geffrye Museum in Clapton, North London. It is assumed that arms, legs and heads were cast individually and dowelled into place using a matching mortar. Of the same approximate age is the statue of Madonna and Child shown in Figure 5.6. This resides (when last heard) in Rutland House, Doncaster, Yorkshire.

Once on display at London's Royal Academy (possibly still so or in store there) is one of Karel Vogel's sculptures meant to be the head of a woman and shown in Figure 5.7. Figure 5.8 illustrates cast stone sculpturing in action where a griffin is being copied using a lump of cast stone. Work shown on these last last two examples would have needed to have been done on cast stone that was not only mature in age but also would have been unlikely to have contained aggregate much in excess of a few millimetres maximum size.

Where cast stone shows one of its advantages over natural stone is when it comes to complex geometries and this is not only seen in the previous figures but also in the Corinthian Cap illustrated in Figure 5.9. This example is c.1948 and was made for 1939–45 blitz damage replacement work in London. The main block was earth-moist cast and the acacia leaves and fruit

Figure 5.5 Cast stone statuary.

Figure 5.7 Cast stone sculptured bust.

Figure 5.6 Cast stone Madonna and Child.

Figure 5.8 Sculpturing a griffin from a block of cast stone.

were vibrated wet-cast into rubber moulds which were peeled off after a few days when the mix had hardened. There would have been no other way to have made these appendages as they have both positive and negative profiles banning the use of a rigid mould. The leaves and fruit were dowelled into holes in the main block and cemented into place using a matching mortar.

Figure 5.10 shows an inn sign manufactured about 1950. It was quite common for cast stone manufacturers to produce name plates, plaques such as that by the *Cutty Sark* on the Thames, and many other bespoke objects.

In a leisure mode, Figures 5.11 and 5.12 illustrate cast stone used in two separate aerator fountains at swimming baths in the London area.

Figure 5.13 shows a c.1935 then the largest cast stone contract ever placed in the Exchange Building in Liverpool. The value of the cast stone supply was £75,000 and, in current day value would have been worth possibly well over a million pounds. Figure 5.14 shows the art deco London Kensington Odeon cinema built 1928–29 which is currently subject to demolition discussion with the intention of building flats from ground level upwards on top of a below-ground multiplex cinema. The façade has a natural granite band course at ground and portal levels and cast stone elsewhere. When mentioned in the December 2006 press, the latest news was that the front façade would be retained.

Figure 5.15 illustrates a c.1990 application of cast stone to a Palladian

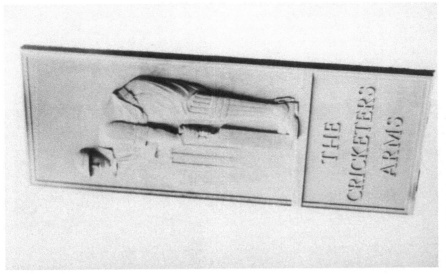

Figure 5.10 Cast stone inn sign.

Figure 5.9 Corinthian Head in cast stone.

Figure 5.11 Cast stone swimming pool fountain at installation.

Figure 5.12 Cast stone swimming pool fountain in use.

Figure 5.13 Cast stone façade to Exchange Building, Liverpool.

Figure 5.14 Cast stone façade to Odeon Cinema, Kensington, London.

Figure 5.15 Modern Palladian-style façade in cast stone.

Figure 5.16 Array of cast stone garden furniture.

façade where the cornices, steps, columns, sills and ashlar were all manu-factured in earth-moist-mix cast stone. Figure 5.16 of the same age, shows the versatility of production in garden furniture manufacture. Some of the background trough units have been 'aged' by a special treatment to produce an old and rustic effect.

Many of these illustrations show that cast stone is not necessarily a product to go into the 'works' as defined by the Construction Products Directive and needs Standard specification accordingly.

6 Machine-intensive processes

6.1 General

As stated earlier, the dividing line between labour- and machine-intensive process has been defined as machine-intensive processes being those requiring a machine which is bespoke to the product in question. This does not mean for products referred to in this chapter that significant numbers of operatives are not involved.

For processes such as vertical pipe production it is known that several operatives are required to deal with attention to the machine and filling, as well as transport across the works and demoulding. On the other hand, it is known that a works manufacturing road paving blocks can have just the one operative at the control station dealing with virtually everything up to stacking and storage in the yard.

Commercial pressure to adopt machine-intensiveness in current labour-intensive processes is all very well and good but the main deciding factor is bespokedness. This is why most precast products being sold from builders' merchants emanate from machine-intensive processes. They are repetitive construction products and not bespoke as a rule. This throws into question the cases of fence posts and similar products which are generally vibrated labour-intensive wet-cast products. The future might possibly see such products being made machine-intensively.

After all, the spiral reinforcing cage that used to be made in the now defunct spun pipe process was produced in an automated machine. A 4-bar cage with stirrups should present no problem to automated production.

6.2 Vibration and pressure

6.2.1 *Vertical pipes*

Pipe diameters can vary from 0.2m to 2.0m and heights up to 4m with wall thicknesses from 25mm upwards. Production rates vary typically within the range 5/hour to 10/hour depending upon the combination of these factors.

In the typical production an inside and an outside mould are assembled at the machine stand. The prepositioning of the moulds is a function of the particular proprietary process and one particular one is described here. The inside mould with high-energy clamp-on vibrators is held in an underground chamber beneath the machine and hydraulically raised when required. (In some processes it is kept above the machine stage and lowered when required.) The base spigot ring is already in place on the machine. The outer mould is placed upon this ring to form an annulus void of the pipe's shape. Where reinforcement is required, rings of steel with their spacers are lowered into the annulus. Some manufacturers drop the reinforcement ring by ring onto the concrete as it is being compacted.

Concrete is fed into the annulus from a cone-shaped feed funnel with intense vibration applied via the inner mould. When filled, the feed cone is removed and a pressurised head ring is applied to the top of the concrete while the vibration continues. In many machines this head ring rotates over a small arc to promote a smooth socket finish. The ring is left in place to help maintain circularity while the unit is moved away from the machine. The ring also gives support to the 'green' concrete during finishing processes. After compaction, the inner mould is drawn back into the underground chamber and the outer mould raised above the machine for the smaller diameters. For the larger diameter pipes the outer mould is sometimes lifted with the pipe and demoulded at the stacking/finishing area. The fork-lift type truck used for moving the pipes from the machine to this area is typically propane gas-powered in order to avoid the possible jerkiness of electric or diesel/petrol-powered vehicles.

A typical mix design would consist of 3 parts m/m of 10–15mm coarse aggregate such as a clean crushed granite or basalt increasing to a 20mm maximum size for thick wall pipes and down to 10mm for thinner wall with 1.5 parts clean sand with 1 part CEM I and a W/C(F) of about 0.4. Admixtures such as workability aids enable the W/C(F) to be reduced to about 0.3, maintaining the same workability. Additives such as PFA are also used as they assist in the compaction process and promote higher green strengths coupled with improved surface finish. Since pipes are exposed to the factory elements and it is essential that a humid draught-free internal environment be maintained. This is specially important when using CEM II and CEM III cements.

Figure 6.1 shows a 'Packerhead' pipe production stage with the operative dispensing concrete into the annulus. Figure 6.2 is of a production line of

Figure 6.1 Filling a Packerhead pipe mould.

Figure 6.2 Demoulding a Packerhead pipe mould.

these pipes with demoulding the outer buckled mould from the unsupported concrete. Figure 6.3 illustrates smaller diameter pipes having been moved from the machine to a storage area. The top pressure rings can be seen on three of these pipes. It may also be observed that the bespoke 'fork lift' truck

Figure 6.3 Transporting a Packerhead pipe across a works area.

is gas-driven. The concrete is only a few minutes old and the operative, standing on a bespoke platform, is carrying out finishing work.

6.2.2 Concrete blocks

This sub-section deals with all masonry units other than autoclaved aerated blocks but includes concrete paving blocks. Concrete blocks, other than paving blocks and specials, are the only products manufactured deliberately to be partially and not fully compacted. The degree of compaction in the mix that has the consistency of earth-moist cast stone is controlled by the depth of penetration of the heads during vibration into multi-staged moulds which commence with being filled with uncompacted concrete. The reasons for this partial compaction are four-fold:

1 better thermal insulation than for fully-compacted concrete. (compare typical conductivity values of 0.3 W/(m·K) with 1.0 W/(m·K);
2 greater production per unit weight of concrete;
3 better bond for render and plaster;
4 improved impact sound resistance in beam and block floor construction.

There are four methods of production, the first three being for the conventional partially-compacted concrete and the fourth for fully-compacted road blocks, also known as 'Interpave'.

Hand compaction

This is a one-off method which is not machine-intensive but has to sit somewhere in the text as virtually the same product is being produced as described in the following two sub-sections. The method is common in many Third World countries where a single mould can be fed from hand-mixed or machine-mixed concrete, hand-compacted and demoulded instantly with following mould re-use. In the UK it has not been observed by the writer for many years but there could be a case for its deployment for specials where a machine modification and a production run would not be viable.

Pallet machine

In this process a multiple block mould box sits under a stationary machine and is fed with concrete which is then trowelled off. The multi-heads descend into the moulds and apply pressure with vibration. The vibration inputs either to the heads or to the mould box itself. The box sits on a pallet and, when the partial compaction is completed, being controlled by the depth of descent of the heads, the box raises and the heads follow and the blocks are left sitting on the pallet with no support. The pallet is slid away from the machine on a conveyor belt or ropeway and a fresh pallet is slid under the machine. The mould box is lowered and the cycle repeated.

The filled pallets may be left to air-cure in the factory or moved into steam or autoclave chambers for accelerated curing and, in the latter case, early dispatch.

Machines can perform anything from a '4-drop' up to a '72-drop' per cycle with each cycle taking of the order of 20 seconds to complete. One can visualise the vast numbers of blocks that can be produced from a large machine in a single shift.

The advantages of the pallet machine process are manifold. The mixer plant, the machine and the curing facility, can all be contained within a relatively small factory and all under cover.

Although much more capital-intensive than the egg-laying process discussed in the next sub-section, it enables production to be continuous largely independent of the external conditions.

Egg-laying machine

These machines are mobile and move along a concrete apron and are generally situated outdoors. The feed hopper, mould box and heads are all contained within the machine. Figure 6.4 shows a typical machine doing a 12-drop of cellular (closed-end hollow) blocks. The principle of compaction is the same as described in the previous sub-section except the ground is the

Figure 6.4 Egg-laying block machine.

pallet and blocks are cured according to the vagaries of the weather. Such machines seen in use in semi-tropical conditions in the Middle East often lose the outside blocks of each drop due to dehydration of the green concrete due to the effects of wind and sun.

The machine hydraulically lowers itself onto the concrete base and remains stationary while the multiple mould is filled and the blocks are made. It then jacks itself up onto wheels and traverses to the next laying position and the cycle is repeated. A cycle can take about 15–20 seconds. When the blocks have hardened, the rows of blocks are scissored together and lifted to storage as can be seen in the background of Figure 6.4.

Apart from the vagaries of the weather, the egg-laying process has several disadvantages:

1   Accelerated curing is not available at the green stage.

2　The machine needs a large area of concrete capable of holding the production as well as being able to withstand the weight and wear and tear of the machine.

3　The ground has to be kept debris-free and clean, otherwise a block laid on a stone could split.

4　The machine requires a large amount of continuously fed concrete to keep it going, with the concrete fed into the hopper by a high-lift dumper-type truck. This means that there would have to be a high speed mixer and a fast truck with efficient communication between the machine and the mixer.

Pallet/egg-laying machine

Although this machine combines the features of the machines described in the two previous sub-sections, it is designed to produce a fully-compacted block. The reason for this is that the process is designed to produce inter-paved blocks for roads, forecourts, garage driveways and similar. Good compaction is necessary for both load-bearing and durability requirements as well as maintaining an acceptable degree of aesthetic attraction under long-term usage.

The machine, illustrated in Figure 6.5 laying a 32-drop of dumbell-shaped units, is mobile while traversing and stationary while compacting with movements being predetermined by photocells and computerised pre-program

Figure 6.5 Pallet-egg-laying block machine.

settings. Blocks are compacted onto a metal pallet plate within the machine with vibration combined with pressure applied via the multi-heads into the concrete-fed metal mould box. The compacted blocks are laid either onto the floor or, by pallet upward indexing at the second and subsequent traverses of the factory floor, onto the previous block layer, by the pallet being slid out from underneath the layer of blocks. Some machines spread a layer of fine sand on top of the previous layer to inhibit sticking between layers. The machine will traverse the block laying factory length about 10 times, resulting in a 'cube' stack of blocks. When the line of stacks has been completed, the machine will traverse sideways and commence, making a parallel set of stacks.

The stacks from these traverses of the machine are clamped and closed together the following day and banded, then taken to storage outdoors for natural curing. The stacks can be steam-cured or autoclaved to accelerate hardening and, for the latter case, expedite delivery.

Figure 6.6 shows dual-square-shaped road block units being made as a 32-drop and Figure 6.7 shows chevron-shaped blocks being laid onto a sand bed which had been spread over a supporting concrete base. The blocks are are pressed into the sand bed with a vibrating plate but that machine is not shown. In the background is the compacting and sand bed levelling float in operation.

Table 6.1 illustrates possible mix designs that could be used for the partially compacted block productions. Readers should not look upon these as being strict starting points as so much depends upon machine,

Figure 6.6 Interpave block manufacture.

Figure 6.7 Block laying.

Table 6.1 v/v typical block mix designs for a 10MPa strength

Type	100 mm solid	150 mm hollow	100 mm fair-faced
Coarse aggregate	5	3	2
Fine aggregate	3	5	4
Cement	1	1	1
Free W/C	0.40	0.40	0.42

ingredients and objectives. Notwithstanding how much one would wish to apply scientific principles to all matters, there will always be a significant amount of alchemy in production. This is known generally as 'know-how'.

6.3 Pressure

As stated earlier, the lines separating the various processes described by their mechanisms are not always distinct. The placement of the hand-compacted block production sub-section with the common machine-intensive processes was an example of this. In this section it will be noted that vibration is sometimes used in conjunction but not at the same time as one of the pressure processes.

6.3.1 Wet-pressed paving slabs, kerbs and edging units

The process consists of having a mix, typically of crushed rock with pulverised fuel ash and, sometimes, natural sand, with a water content about twice the level of the target finished product and pushing the excess water out of the concrete through filters using hydraulic pressure. The commonly used press turntable is generally circular in shape and consists of three stages, each with a fixed mould in it situated at 120 degree intervals. As each mould stage and the press head have to be changed for changes in production for 'specials', there is also a rectangular-shaped reciprocating mould table with a single stage mould.

Describing the three-stage machine, each mould has a perforated steel base plate which sits on top of an ejector jack. A sheet of paper is then placed upon the perforated plate. The high water content mix is accurately dispensed from a secondary hopper onto the paper as illustrated in Figure 6.8 in a rotary press and in Figure 6.9 in a reciprocating press. The bottom face of the product as-cast is normally the top (paving flag) or side (kerb or edging) face as used. Sometimes another sheet of paper is placed on top to assist top filtration. The table turns 120 degrees and a hollow perforated face press head, with the hollow kept under vacuum, descends into the mould, applying a pressure of the order of 400T. The excess water is pressed out through the bottom paper and sucked out through the hollow and perforated face pressure head. The table then rotates a further 120 degrees, the product is jacked up out of the mould box and the top paper is removed

Figure 6.8 Wet-hydraulically-pressed slab production-rotary table.

Figure 6.9 Wet-hydraulically-pressed production-reciprocating table.

sometimes by blowing it off and a vacuum lifting plate shown in Figure 6.10 picks the unit up. The bottom paper is removed, often by blowing with small compressed air jets, the lifting plate rotates into the vertical position and places the product on its edge on a pallet. Sometimes the bottom paper is left in place until the concrete has hardened the following day but the paper has to be of a low size content (aluminium sulfate) quality to avoid sticking due to gypsum formation at the paper/concrete interface.

In the reciprocating press the machine is, effectively, single stage as the mould moves sideways to have the pressed product ejected and vacuum lifted followed by refilling with the mix and slid under the press head. The advantage of the reciprocating machine is when uncommon geometries need manufacturing as there is only one mould and press head that need changing.

Each stage of the rotary machine is in continual use. In effect, as a product is being pressed another product is being demoulded and a third stage is being filled.

Mention was made earlier of the use of a crushed rock aggregate. The green strength property requirement needs aggregate interlock to take place and this precludes the use of natural gravels unless they are used as crushed aggregate. One can imagine what would happen to a 600mm × 600mm × 50mm paving slab stood on its edge on a pallet that is to be moved at a few seconds concrete age unless there was a high degree of rigidity in the pressed material

Cycle times vary widely depending upon a large range of factors but the

Figure 6.10 Vacuum lifting wet-hydraulically-pressed product-age zero.

press part of the process is the limiting factor as it is the slowest part of the cycle. Typically pressure is built up in about 5 seconds, the top pressure is held for about 10 seconds and released over 1 second. In effect, the typical production rate from a 3-stage rotary press is about 3 units per minute.

Tables 6.2 and 6.3 show some typical fine aggregate gradings as well as mix designs.

6.3.2 Earth-moist pressed paving

The rotary press table is typically a four-stage process but sometimes can be a five-stage and the machine is used for paving slabs which tend to be 'specials'. The products often tend towards the architectural market rather than the civil engineering application discussed in Section 6.3.1. The mould base plate, instead of being a perforated plate, is generally made of solid steel or a

Table 6.2 Fine aggregate gradings for wet-pressed production

Passing sieve no.	Crushed rock	Natural sand
5	70–100	90–100
2.36	50–100	80–100
1.18	40–90	70–90
600	30–80	50–90
300	25–50	20–60
150	15–30	10–40
75	5–15	5–15

Table 6.3 Some typical mix designs for wet-pressed products

Aggregate	Kerb	Kerb	Slab	Slab
20 mm	1.5	2.0		
6 or 10 mm	2.0	2.0	2.0	2.5
5 mm down	2.5	2.5	2.5	2.5
Natural sand	0.5		0.5	
PFA		0.5		0.5
Initial free W/C	0.80	0.86	0.70	0.80
Final free W/C	0.40	0.43	0.35	0.40

very hard rubber. Underneath each plate is an ejector jack as for the wet press process.

As an example of a hexagonal-shaped architecturally faced paving slab all five stages would be put to use continuously as follows:

1   A workable facing mix would be placed in the mould.
2   The table rotates 72 degrees and vibration is applied.
3   A further rotation takes place and an earth-moist backing mix is discharged onto the compacted facing mix.
4   Rotation to the press stage then takes place and a press head descends into the mould, applying a pressure of several hundred tonnes, causing the earth-moist mix to compact and, at the same time, promotes inter-mix diffusion and bonding to occur with some of the water from the facing mix being drawn into the backing mix.
5   The table rotates to the demoulding stage where the ejection jack raises the base plate and tilts it to the extent of sliding (with side pushing if required) the product off onto a holding pallet where the unit is stored on its edge. The backing mix texture is too open to permit vacuum lift-off as is possible for wet-pressed products.

A variety of non-standard shapes such as square, rectangular, triangular, hexagonal, octagonal, semi-circular and circular can be manufactured although the last shape would have to be stored either flat or on special edge-holding pallets. Not only these shapes but special exposed aggregate panels can be made in the form of three-drop rectangular units shown in Figure 6.11 being aggregate exposed immediately on leaving the machine.

Mixes are typically of crushed rock together with natural sand, with the sand often required to be more rounded than for other pressed processes. Obviously, for faced pavings, special mixes need to be used and these need to suit architectural specifications.

Aggregate gradings tend to be finer than those deployed in the wet press process and Table 6.4 shows typical gradings for crushed rock and natural sands and Table 6.5 shows mix designs that could be used for these machines. The same comment about 'know-how' made earlier equally applics.

Figure 6.11 Exposing aggregate on wet-hydraulically-pressed slabs.

Table 6.4 Fine aggregate typical gradings for earth-moist pressed products

Passing sieve no.	Crushed rock	Natural sand
5	70–100	90–100
2.36	50–90	80–100
1.18	40–60	70–90
600	30–50	50–80
300	20–30	20–60
150	0–10	10–30
75	0–5	0–10

Table 6.5 Suitable earth-moist mix designs for pressed slabs and bricks

Aggregate	Slab	Brick
6 mm crushed rock	3.0	–
Natural sand or rock sand	1.5	10.0
Free W/C	0.32	0.45

6.3.3 Wet/earth-moist cast floor tiles

The process is similar to the production of a faced paving slab refered to in Section 6.3.2 but a multi-stage turntable with 4–8 stages is used with only four stages being deployed in the process. The bottom of each mould, typically 300mm square by 25mm deep, has a removable base plate acting as a demoulding pallet under which sits an ejector jack. The process consists of a combination of pressure with water inter-diffusion and mix interlocking between a backing and a facing mix. The five stages typically consist of:

1   An earth-moist mortar backing mix is placed in the mould base.
2   A very workable cement/water mix is placed on top.
3   The coarse aggregate is hand-placed into the cement slurry.
4   A press head descends and applies a pressure of 1–2 tonnes.
5   The pressed tile is raised by the ejector jack and manually placed onto a larger pallet to harden.

The process is both machine- and labour-intensive as there are parts of the production that a machine would find difficult or impossible to handle. The word 'concrete' for the facing mix was so written to emphasise that a typical terrazzo tile could have a maximum aggregate size of 40mm. To deal with getting a 40mm aggregate size concrete into a depth of 10–15mm of mould means that a high flakiness index aggregate has to be specially selected and the aggregate has to be spread by hand quickly before the table rotates to the press stage.

   Once hardened the following day, the tiles are de-palleted and run through several grinding and polishing phases to produce the well-known terrazzo tile finish. Polishing used to be done by diamond grinding wheels but is now changing to the use of laser cutting and polishing devices.

6.3.4 Brick production

It is not known if concrete brick production is still in operation in the UK but the process is nominally the same as the pallet machine with a similar multi-drop. The bricks are fully compacted and can be frogged if required.

6.4 Extruded

Since all these processes rely upon the principle of the chef's icing bag, they only require a pressure application and a product cross-section-shaped orifice to push the mix through to form the requisite shape. In precast production there are two main sub-processes: one deals with mortar as used in roofing tile production and the other with concrete for the general production of pretensioned prestressed floor planks, beams, columns, wall panels and similar. The division between extrusion and slip form production (as applied to in situ concrete) is, again, rather tenuous, so they are discussed together in Section 6.4.2. There is a similarity between precast concrete extrusion processes and pumped in situ concrete in that in both cases the mix is put under pressure which is relieved as it passes through the die or the pump nozzle. Both precast and in situ processes are very sensitive to both coarse and fine aggregate gradings as well as overall mix design. The particular range of acceptable mix designs for a specific process and product has to be established by experimentation with a lot of concrete or mortar wasted before getting things right.

6.4.1 Roofing tiles

A high cement content low water cement ratio mortar mix is fed into a hopper, at the base of which is situated a relatively low speed rotating vaned roller forcing the partially-compacted mix into a tapering tunnel, at the end of which is a high speed vaned roller pushing the now fully-compacted mortar through a tile cross-section orifice. This ribbon of tile section-shaped mortar is fed onto a continuous ropeway train of oiled aluminium pallets, each having the profile of the underside of the tile together with the manufacturer's name. A guillotine cuts the ribbon at pallet joints and, if required, coloured sand is dispensed onto the surface with the excess being blown off by compressed air. The tiles are ropeway conveyed and automatically stacked and placed inside steam-curing chambers.

After curing, usually the following day, the tiles on their pallets are automatically fed back onto the ropeway and passed between blades operating at the mortar pallet interface which demoulds the tile. The tiles travel in one direction to the stacking area and the pallets are brushed and oiled and fed into the stack that feeds the machine.

Duplex machines can extrude two tiles per second at a time, side by side, and if these tiles were large roof slates weighing nearly 10kg each and the machine extrudes these on a cycle per second basis, it can be seen that roofing tile production is a large user of cement. As an approximate calculation the 20kg/second tile mortar production could equate to 5kg cement/second.

6.4.2 *Extruded planks, beams and panels*

There are two main methods of production with the principle being the same in both cases. Both use a machine that moves along a prepared bed consisting of a metal base with all the required prestressed wires and strands laid along the bed passing through the machine and stressed to the design levels. The machine is self-propelling with the force arising from the augering strength of the concrete extrusion pressure.

In the first type of process a number of augers in the machine under the feed box push the concrete through a rectangular die of the dimensions and geometry of the product cross-section. The concrete is compacted within the space between the augers and the die and within this space is contained the prestressing wires/strands. Further compaction of the top surface is sometimes obtained by high frequency clamp-on vibrators on the trailing roof of the machine. The resulting product is a continuous ribbon of concrete with nominally circular holes where the augers were situated. Figure 6.12 shows a stripped down three-hole plank machine with augers visible inside the tubes.

A similar process is used in the aggregate auger machine, the only difference being that both concrete and loose aggregate in the void shape are extruded together. The aggregate is recycled. This second type of extrusion has the advantage of producing voids other than circular in shape. Figure 6.13 shows a close-up of the section of a prestressed unit and Figure 6.14 illustrates the application to the production of prestressed ribbed wall panels.

Figure 6.12 Extrusion machine box for prestressed slab/beam production.

Figure 6.13 Extrusion production using loose aggregate for void-forming.

Figure 6.14 Wall panels produced by extrusion.

Mixes are commonly of flint gravel, coarse together with a fairly fine concreting sand. As stated earlier the process is very sensitive to aggregate gradings which, if not well controlled, can result in inadequate compaction of the top layers of the concrete. The coarse and fine aggregate shapes are also important. A rounded gravel and sand will favour optimum extrusion but will be a detriment to the green strength. Therefore a suitable medium has to be struck in the selection.

Curing is often done by covering a line of production with polythene or similar and heating with perforated steam pipes under the beds but other methods can be used including heating the mix water. When the concrete has reached the required strength, the end-bed prestress is released and the ribbon of hardened prestressed concrete is cut into the required lengths.

6.5 Autoclaved

This process mainly applies to autoclaved aerated concrete blocks (AAC) but the process is also used for conventional concrete block although the word 'concrete' does not accurately define the hardening mechanism. However, in texts and Standards, autoclaved blocks are grouped with all other masonry units including calcium silicate products. Reinforced auto-claved aerated cladding panels used to be made in the UK but are only thought to be in current production in mainland Europe.

The difference between autoclaving and other types of heat curing is that autoclave curing is carried out at about 200°C temperature and 12 atmospheres of pressure. This results in a chemical reaction between any suitable lime and fine siliceous components to form calcium silicate, not the calcium silicate hydrates obtained from cement hydration. In effect, once the product is autoclaved, further curing is not possible and the products can be dispatched to site without delay. This is why it is not uncommon to see blocks, especially AAC, arriving on site in a hot and steaming condition.

Since the ordinary block production has been described in Section 6.2.2, only the AAC process will be described.

The mix for AAC production is typically a blend of PFA with Portland cement or fine ground silica sand with hydrated lime. Each mix also contains aluminium powder and other chemical admixtures. Heated water is added to the mixer and the slurry is poured into a mould of a typical size 1m wide by 1.5m high by 3m long.

The aluminium powder reacts with water to form hydrogen gas bubbles, causing the mix to rise as an aerated mortar mass. This mass is allowed to take on a little green strength and the excess overspill is removed. The block has its side and end plates released and is then cut in all three orthogonal directions by wires to leave a multi-block master block. Figure 6.15 shows the master block with the mould sides removed going through one of the three cutting stages. Figure 6.16 shows the master blocks steaming on their removal from the autoclaves which can be seen in the background. The

Figure 6.15 Expanded foamed mortar block being cut before autoclaving.

Figure 6.16 Completed master blocks removed from the autoclave.

doors of these autoclaves, although they appear to be circular, are not. They are elliptical in shape and are locked by being turned into place. The process has to have many safety precautions as the pressure on the door is several hundred tonnes.

7 Accelerated curing

7.1 General

Several decades of experience by the author have been condensed into this one chapter. It reflects only on site visits and research under the auspices of the British Precast Concrete Federation and Laing Technology (previously Laing R&D) and does not claim to cover all possible methods. It deals with one method (electrical induction) that came and went in the UK and which perhaps merits reconsideration.

The process of cement hydration is exothermic with a small initial output of heat as soon as water comes into contact with cement due to physical processes.[49] This is followed by a longer period of chemical exotherm as water reacts with the cementitious components. Like most chemical processes, the warmer a system becomes, the faster is the chemical hydration. Thus, a concrete mix starting off at 5°C will warm up and gain strength slowly whereas a mix starting at 20°C will warm up more quickly with an accompanying faster gain in strength. Taking an example of an accelerated mix starting at, say, 40°C, this product could possibly be handled and stacked after a few hours. Most methods of heat acceleration curing generally detract slightly from the 28-day compressive strength, a common criterion used for assessing concrete. The detraction in strength is often just in the range of being significant but the vast majority of precast processes have plenty of 28-day compressive strength in hand as the important thing for precast production is the flexural strength at the demoulding age which could typically be from a few hours to a day old.

The discussion that follows gives general information on matters under consideration in accelerated curing. Since the spectrum of products made by the precast industry is huge, each particular production needs to be assessed within its own variables such as materials, geometries, heating costs, demoulding age, storage, and so on.

Before going into a discussion of materials there are two aspects of curing which often tend to be considered by the parties involved as being separate entities. The word 'curing' is anomalous as one needs to define or explain what sort of curing is under consideration:

1   curing to retain heat and/or control heating/cooling rates;
2   curing to prevent too rapid a loss of moisture.

From the heat point of view, it would seem that the more exotherm heat that can be retained, the faster the turnround and re-use of the mould. However, this needs to be balanced by the factory or site considerations. A larger section of concrete demoulded at, say, 50°C average temperature and immediately exposed to, say, 10°C on a cold windy day, could suffer excess strain due to excessive thermal differentials. These could result in cracking or other distress. Therefore, one consideration is the control of the cooling cycle as well as the demoulding regime as the concrete hardens. Virtually full licence is practical in the heating cycle as the concrete is fresh and relatively plastic and can take high thermal gradients.

On the second point above, moisure retention, including a controlled rate of release of water that is excess to the hydration requirements, is an important aspect. Surface and in-depth cracking and dusting can occur if this is not kept under control. Moisture movement in the surface layers due to too fast a rate of water loss can lead to crazing and/or drying shrinkage cracking.

What these two previous paragraphs emphasise is that several curing variables have to be kept under control at the same time and concentration on one at the expense of the others will only result in problems. It has to be borne in mind that whatever the state of the concrete, fresh, green, hardening or hardened, one is dealing with a multi-variable system. Control of one variable, ignoring the others, can only be descibed as unwise.

7.2 Materials

7.2.1 Moulds

The mould material plays an important part in the selection of the accelerated curing method to be used. Steel is several hundred times more thermally conductive than timber and is used in much thinner sections than timber. A wooden mould 25mm thick has a similar thermal resistance to a 500mm thick brick or 10m thick steel wall. Therefore, the application of an external heat source such as atmospheric steam or using a heated enclosure would not be viable for timber moulds. In addition, the behaviour of timber under hot damp conditions will cause warping and distortion. On the other hand, the use of heated concrete in a timber mould would be more attractive than for steel or GRP mould due to the thinness of the latter type of moulds due to their large radiative properties. Their enhanced heat loss due to conduction and radiation also helps in controlling cooling rates by adjusting the surrounding atmosphere and/or insulation that one could have around the mould. Thermal insulation of steel or other thin section mould materials can be improved by lining the external surfaces with a low conductivity material such as expanded polystyrene. Radiation loss from steel and plastics moulds can also be reduced considerably by painting the exterior surfaces with a low heat emissivity material such as aluminised paint. Expanded plastics can also be used to cover all mould exposed concrete faces where its properties, both to conserve heat as well as to inhibit too fast a moisture loss, can be utilised.

Hollow steel moulds can be used for heat curing by passing heated air or steam or other fluid through the cavity, taking precautions to minimise heat loss through exterior surfaces. This method was used successfully in precast production during the post-war housing construction boom. The author has no knowledge as to whether or not it is in current UK use.

Care needs to be exercised where the geometry of the concrete is other than that of a rectangular parallelapiped. Examples of these variations include ribs, returns, window and doorway openings. As the concrete heats up in the plastic state, stresses are unlikely at these change of geometry areas but, during the cooling stage, the concrete is hard and has lost much of its ultimate strain capacity.

7.2.2 Insulants

Although the types of insulant material available have not changed much since the first edition of this book, the appreciation of their properties has increased and their use in both precast and in situ concrete work is now common. Their value in thermal insulation can be appreciated by comparing thermal conductivity values in $W/(m \cdot K)$ with other known materials:

concrete (density $2350 kg/m^3$)	1.3
steel (mild or high tensile)	50
expanded polystyrene (EPS)	0.035
softwood timber	0.13

In accelerated curing applications expanded polystyrene is the most common material used as it is cost effective, easily cut and easily glued, as well as being recyclable. It can be used as mould lining, covering the outsides of steel and similar moulds and covering the fresh concrete exposed faces. Expanded polystyrene hollow interlocking blocks are used as permanent formwork to produce insulated in situ concrete walls which can be used in single or inner leaf cavity external wall construction as well as for internal walls.

When used in sandwich panel construction, the purpose of the EPS is to improve the thermal properties of the precast unit in its deployment on site. However, when used as a sandwich which does not span the complete section (in effect, a part sandwich), it promotes accelerated curing over that area and this can often be seen on the visual face where different colour concretes occur side by side due to the differences in the curing conditions.

The use of insulants in moisture curing application is possibly not so much appreciated but is still of benefit especially for low humidity environment production. It is stressed that although cold air cannot hold as much moisture as warm air, the cold air still has the capacity to desiccate the surface of concrete enough to cause distress.

7.2.3 Release agents

In addition to normal curing temperature property requirements demanded of release agents (as well as those of surface retarders) these same properties need to be retained at elevated temperatures as well as at the extremes of humidity. Generally speaking, grouping these agents into classes, the chemical release agents and mineral oils perform the most satisfactorily. Mineral oils tend to vary from manufacturer to manufacturer and precasters should be prepared to ring the changes and not to denigrate them as a class because a specific one performs badly.

At elevated temperatures, chemical release agents do not release significant quantities of volatiles whereas volatiles could be released from mineral oils. Good ventilation will inhibit this nuisance but it can be a dermatitic and/or allergic hazard for operatives having this sensitivity or having a family history of problems.

Retarding release agents are usually used to produce exposed aggregate finishes and can be water-based or organic solvent-based systems. The organic solvent-based retarders have been found to perform best for accelerated curing applications. Evaporation of the solvent leaves a hard waxey film on the mould surface that is activated by contact with alkaline systems such as the lime in cement.

7.2.4 Concrete ingredients

Aggregates

Aggregates used in concrete may be natural or synthetic or mixtures thereof and the selection is outside the remit of this book. What needs to be borne in mind is that the wide range in physical properties of what might be used plays a significant role in the behaviour of concrete during both normal and accelerated curing. Three main factors under consideration are density, specific heat and the coefficient of thermal expansion. Specific heat at about 0.2 cal/g (0.84 J/g or kJ/kg) is the only one of these three that has approximately the same value for all aggregates. Where it plays an important role is in the effect of exotherm on various mixes.

Take as an example a comparison of two concretes both with a cement content of 350 kg/m^3 but one with 2000 kg/m^3 of granite aggregate and natural sand and 150 kg/m^3 of water and the other with 1500kg/m^3 pumice aggregate and natural sand and 180 kg/m^3 water. Assume both concretes start off at 10°C and the cement gives off 20 kcal of heat during the first 24 hours. To find the mix temperature at 24 hours old, the following equations assuming adiabatic conditions may be used:

Final temperatures : granite mix T1, pumice mix T2

2.35(T1 − 10) × 0.2 + 0.15 × 1(sp.ht of water)= 20 Giving T1 = 50°C

1.85(T2 − 10) × 0.2 + 0.18 × 1(sp.ht of water)= 20 Giving T2 = 60°C

These equations are very approximate as they have assumed that the water remains in its initial form and ignores the value of the specific heat of the wet 24 hour-old concrete. The equations illustrate the significant effect that low density aggregates can have on heat gain but it needs to be emphasised that the calculations are approximate. A concrete starting off at 20°C will give out heat much faster than one starting at 10°C temperature and, in turn, will accelerate the chemical reaction.

The coefficient of thermal expansion of the granite concrete will be about 1.5 × 10EXP−5 whereas the pumice concrete will be about a third of this. This means that less thermal shrinkage will take place during the cooling cycle with the pumice concrete than with the granite aggregate concrete. This not only results in less strain occuring at ribs and openings in the mould but the lightweight aggregate concrete will have a lower modulus and higher ultimate strain capacity than the granite concrete, both factors contributing to a lower risk of distress.

Before going onto the next sub-section perhaps a few words about cracking will not go amiss. Cracking in many building materials, both natural and synthetic, is common; it depends upon what is visible and what is not. Panic buttons in the concrete world tend to get pressed when they are visible to the naked eye. However, what the eye sees is the crack at the surface so that all that one can record is the 'surface crack width' and not the 'crack width'. Without further knowledge one has no idea of the following:

1   Is the crack a single crack as it descends or does it fork out into two or more cracks?

2   If it does change to multi-cracks, are these joined to the lead crack?
3   If the crack remains as a single crack, is its width constant or does it taper or widen as it descends into the concrete?
4   How deep into the concrete does the crack penetrate and does it constitute a corrosion risk if touching the steel (for 'crack' read also 'cracks')?
5   Is the quality of the concrete more of a corrosion risk than from ingress of fluids through the crack(s)?

Cements

There have been many changes in cements and additives over the past couple of decades,[50,51] but the Standards are not helpful when it comes to how these changes relate to accelerated curing processes. The finer (higher specific surface) a cement becomes, the more rapidly it reacts with water, resulting in accelerated setting and hardening times and an increased rate of exotherm. Reference was made early in this chapter to the effect of increasing the starting temperature of the mix in that the warmer the start, the more heat is produced and the more rapid the reaction. This situation can lead to a loss of control, especially for large sections of precast units where excessive thermal gradients can be generated.

Additives such as PFA and GGBS have been used in precast manufacture for over 50 years and are now available as cement blends. The pozzolanic reaction of PFA with cement and the secondary hydration reaction of GGBS have been shown to be of great benefit in accelerated heat curing processes. Again, it is emphasised that all variables in production need to be assessed under works conditions to ascertain the best combination of ingredients and curing cycle.

Admixtures

These are not considered to be of significant application for concretes where accelerated setting and hardening rates are required by the use of heat. It is emphasised that because a particular plasticiser or superplasticiser allows a specific reduction in water content for the same maintained workability at normal temperatures that the same reduction will apply at elevated temperatures. Each application should be assessed under as near works-identical conditions as possible to the actual usage.

7.3 Curing methods

The methods outlined below group the various ways of using heat to accelerate curing into convenient sub-sections. These cover most of the applications while, at the same time, highlighting one method that came and went in the UK industry, namely electrical induction. It may well be in the years to come that energy will become more attractive as an electrical supply, therefore one's options should be kept open. It also needs to be borne in mind that electricity bought off-peak can result in attractive energy costs whereas the same thing cannot be said for oil or necessarily for gas. Which of the methods is selected depends upon what is being made, the type of mould being used, the number of castings required, the turnover target rate and the energy cost. It also needs to be kept in mind that the use of the heat generated by cementitious hydration exotherm is possibly not used to its best advantage.

It matters little which method is selected unless two important properties of the system are known, monitored and controlled:

1   the concrete temperature and temperature gradients, the latter being of special importance when the concrete is hardening;
2   relative humidity, temperature and, if outdoors, the wind speed at the exposed concrete faces.

The temperature of the concrete relates to the early strength and to the likelihood of stress cracking due to too high a temperature gradient. As far as relative humidity, temperature and wind speed are concerned, too drying a condition can promote surface crazing, cracking or dusting and too high a level can lead to surface puddling, softness and lime bloom (often mis-named as efflorescence).

7.3.1 Steam atmosphere

Although called 'steam', the medium used is hot water vapour at an ambient temperature below the boiling point of water, namely 100°C. Being one of the currently most viable method capital cost-wise, it is probably the most popular. Sometimes the steam may be recirculated after re-cycling the condensate but more often than not it is wasted to the atmosphere.

The steam can be:

1   passed through openings into thermally-insulated steam-curing chambers, a battery of which is illustrated in Figure 7.1;
2   passed through pipes with openings (sparge pipes) situated under the mould beds which are then covered with cloches;
3   passed through hollow steel moulds;
4   passed through underfloor heating pipes heating the mould bases by metal plate contact or similar.

The common method of producing steam is by a steam boiler but an oil heater feeding pipe laid in water troughs under the mould beds is an alternative way of generating steam involving higher capital costs but far less maintenance.

The processes described in (1) and (2) need to have relative humidity

Figure 7.1 Battery of steam-curing chambers.

control in the range of 75–90 per cent rh as exceeding 90 per cent rh promotes roof condensation and dripping onto exposed faces which can result in marking, cratering and/or puddling. Too low a humidity can lead to excessive drying with possible surface weakness and/or dusting. Methods (3) and (4) will be with the moulds covered within which areas humidities can build up to too high a level and therefore venting may be necessary.

7.3.2 Steam injection into the mix

This time, the steam being superheated a little above 100°C, is true steam and conveys heat to the mix by two mechanisms. The first is the specific heat transferred as the steam cools the few degrees to its boiling point and the second, the main one, is the latent heat emission as the steam condenses into water. The method needs very strict control as the steam-added water means that the mix needs to be below its design water content as steam condensate water is added during the injection. Thus, without the required control, one could end up with too workable a mix at the target temperature or too hot a mix at an unacceptable low workability. Applying this control it has been found that a mix at 60–80°C can be produced in 5 minutes. Should the requisite control not be applied, it has been found that mixing times up to 20 minutes can become necessary. This would be unlikely to be a viable process.

The method originated in Denmark several decades ago and was found to be most useful for large section precast units such as bridge beams. Smaller section units were observed being subjected to additional accelerated curing using conventional steam.

Conventional pan mixers can be modified to take steam injection but proprietary mixers are available.

7.3.3 Aggregate heating

Either by 'underfloor' or steam pipe heating this method is useful during cold weather deliveries of washed aggregates as ice can form which, if used in the mix, can cause detrimental effects. Again, as for steam injection, the aggregate stock pile will take on a higher moisture content due to condensation. Although this will not be to the same increased level as for steam injection, allowance needs to be made for the change in the water required at the mixer stage.

Aggregate, like cement, has a specific heat about one-fifth of that of water and therefore a large quantity of heat is needed to heat it. However, whichever heating method is used, the heat input has to be shared by a number of materials in the mix and the aggregate content will have the largest heat sink capacity, more than any of the other mix ingredients.

7.3.4 Electrical 'immersion' heating

This method simply involves left-in-concrete insulated heating wires through which is passed a low voltage high amperage current. Figure 7.2

Figure 7.2 Cast-in electrical heating wires.

illustrates two runs of spiral heating wires placed beneath the reinforcing mesh. The spiral shape of the heating elements promotes optimum heat dissipation. The concrete temperature needs to be monitored so as to avoid over- or under-heating and heating coils need to be connected to an electrical control unit. In addition, the coils need to be electrically insulated to avoid shorting out to the reinforcement and/or the mould while, at the same time, having a minimal thermal insulation property.

7.3.5 Electrical mat heating

This method uses electrically-heated mats fixed to the mould sides with suitable thermally conductive electrical insulation covering them. Again the method needs to have a monitoring and control system to control temperature. Use of the mat approach has the advantage of being able to apply different mats to different parts of the mould where changes in section would require different levels of heat input.

7.3.6 Electrical induction

This method relies upon direct electrical current transmission through the concrete itself. It was used in general production of precast units at a UK factory for some time but, for reasons unknown, its application was terminated. Its main disadvantage is that it is capital-intensive in both circuitry and moulds. Its main advantage is that to a large extent the concrete itself controls the heating. The reason for this is that fresh concrete is much more electrically-conducting than in its setting or hardening stages. In effect, for a constant DC voltage supply the current will decrease as the concrete gets both hotter and harder as its electrical resistance increases.

Figure 7.3 shows the complex circuitry control and Figure 7.4 a bed of railway sleepers being cured by electrical induction.

7.3.7 Hot mixing water

Placing this last in this list is no reflection on its preference as the use of hot mixing water only requires a large thermally-insulated hot water tank storage and a method of heating that water. If electricity is used, 'white meter' overnight power can be attractive and, in some areas, off-peak gas can be used to economise. Water has about 5 times the specific heat of the other mix ingredients and is easy to heat and store. However, in weight, it only constitutes about one-fifteenth of the total mass of low specific heat materials. This approximately means that heating water from 10°C to 70°C will for cause a $60/15=4K$ rise in temperature. This ignores the thermal capacity in the mixer plant whose thermal inertia can be significantly overcome by starting the first shift by pouring a litre or so of methylated spirits into the pan and igniting it. Petrol or diesel should not be used as igniting these chemicals will

Figure 7.3 Control circuitry for induction heat-curing.

Source: Reproduced with the courtesy of Whittles Publishing Ltd

leave a residual film of black carbon. Great care should be exercised in this activity and the company's Health and Safety Officer should be consulted before doing this.

7.4 The curing cycle

In order to avoid confusion between temperature and temperature change (rise or fall), change is shown by the letter 'K' (see Section 7.3.7) where actual temperature is shown by 'C'. This follows the practice of unit description used in European and British Standards.

Examples of curing cycles given in the first edition have been repeated here but they are for guidance and are not recipes. The reason for this comment is that additives and admixtures are now in much greater use than they were in the early 1980s. Additional to this, the composition of Portland cement has changed significantly in the ratios of the main calcium silicates. The earlier cements had a predominance of C2S over C3S but this has reversed over the past two or three decades, mainly because of the aim of achieving faster hardening and higher earlier strengths and quicker turnaround for both precast and in situ concretes. C3S is much faster in hydrating than C2S.

Figure 7.4 Electrical induction curing of railway sleepers.

Source: Reproduced with the courtesy of Whittles Publishing Ltd

Therefore, it is essential to establish what curing process and regime suit one's specific product and, just as importantly, what range one can work within safely. This particular sub-section discusses the variables that affect the process and Section 7.6 summarises all the foregoing in general guidelines.

7.4.1 The heating and holding cycles

The main pitfalls that await the unwary are:

1   too high a temperature gradient;
2   too high a temperature;
3   too low a temperature;
4   too fast a loss of water from the surface.

Fresh and hardening concrete can stand a lot of abuse as differential thermal and/or moisture gradients can be tolerated without leading to distress in the form of cracking. A body of opinion used to support the theory that an ambient room temperature hold time of 1–3 hours before applying heat benefited the ultimate strength but extensive testing found that there was no evidence to support this theory. Common sense dictates that it is best to apply heat as early as possible after compaction. Late application of heat, especially if applied when hardening has commenced, could cause cracking.

Heating rates should not exceed 20K/hour taken at the position of the maximum concrete temperature. This should be reduced to 15K/hour for lightweight aggregate concrete (although not in common current usage) as its lower thermal conductivity can result in 'soft centre' formation. Once enough heating has been applied, the heating source should be either switched off or moderated to control the cooling cycle. As the cement exotherm continues after this cessation or moderation, it can be found that this time is adjudged when the maximum concrete temperature is of the order of 5–15K below the target maximum. Which end of this range is targeted largely depends upon the thickness of the section as there will be continuing production of exotherm heat from the cement. A thick section (viz. 300mm × 300mm) would need to be at the 5K end of the range whereas a thinner section (viz. 100mm × 100mm) nearer the 15K end.

The target maximum concrete temperature has been the subject of much argument. Provided that precautions are stringently taken concerning temperature stresses and both moisture and temperature curing, it has been found that curing up to 90°C concrete maximum temperature is possible without problems arising. It is assumed that health and safety aspects would be observed and insulated gloves worn during handling production.

This 90°C maximum would need to be reduced for units with window and door openings as well as other restrictive changes in geometry. From the practical and economic points of view, the amount of heat added would relate to the early strength requirement and 50°C–70°C would be the typical industry's range of maximum concrete temperatures.

The optimum holding time is also part of the heating cycle and is a function on the target early handling strength and the planned turnover cycle. For a typical 24-hour cycle, a unit would possibly require demoulding and handling at 16–20 hours old and a holding time at its maximum temperature of 7–10 hours following, say, a 4-hour-long heating cycle, could well be of the right order. As a general rule, the maximum holding time makes up about half the total curing cycle time with the total cycle including a significant part of the cooling cycle.

7.4.2 The cooling cycle

This is the most critical part of the process as the concrete has hardened and has a restricted toleration to strains arising from excess thermal and/or moisture gradients. Again, common sense dictates that it would be extremely unwise to take a concrete at 80°C and demould it and place it outdoors in an ambient temperature of, say, 10°C. Cooling rates, depending upon the thickness and shape of the section, should generally lie within the range 10–20K/hour and temperature gradients, again depending upon the thickness of the section, restricted within a range of 50–150K/metre. Again, the main consideration is the geometry of the unit.

In estimating the curing maturity, a measure can be taken from the area under the temperature/time curve in the program used. This can be based upon 0°C as the baseline rather than the −10°C used for in situ concrete maturity prediction.

Table 7.1 gives guidelines on some of the various conditions that could obtain and can be used as starting points in any planned curing regime. It is stressed that this table does not cover mixes containing additives nor does it take into account the behaviour of admixtures at elevated temperatures. Precasters are advised in all cases to assess their particular conditions in order to establish optimum ranges for all the variables previously discussed.

7.4.3 Temperature and moisture curing

The maximum temperature and moisture gradients one can tolerate for a specific production will need to be established by experimentation working towards the extremes where distress occurs. For both gradients this will establish critical levels and lead to the setting up of safety limits within which one may work.

Temperature excesses may be avoided by the use of insulated moulds and/ or environment control, and moisture excesses (too much or too little loss) by the use of insulation materials, membrane coverings and, possibly, the use of vapour control admixtures. For concretes placed within steam curing chambers, too low a relative humidity (rh) will lead to surface drying, crazing, cracking and/or dusting and too high an rh leads to puddling and staining at the exposed surfaces. In addition to these latter risks and as previously remarked, condensate forming on the soffit of the chamber or underside of the covering can drip onto the concrete. It has been generally found that maintaining the rh in the range 75–90 per cent will suit most processes.

Guidance on surface moisture gradients is difficult to give quantitatively as so much depends on a multi-variable situation. Suffice it to say that a compromise has to be achieved in inhibiting too fast a rate of emission of free water coupled with the need for the concrete to dry naturally. A visual assessment is possibly the best way of judging this with the aim of keeping in between light-coloured dry surfaces and sheeting with a water film. It is best to lean towards the latter situation where possible.

7.5 Monitoring and control

Cast-in disposable thermocouples are the most viable method of monitoring temperatures and temperature gradients and thermocouples can be located at selected positions which would include zones at most risk from excesses. With modern control equipment these thermocouple outputs can form part

Table 7.1 Curing cycle comparisons

Start temp. (°C)	Heat (h)	Temp. (°C)	Hold (h)	Hold temp. (°C)	Cool (h)	Final temp. (°C)	Maturity, heat (°C h)	Maturity, hold (°C h)	Maturity, cool (°C h)	Maturity to 24 h (°C h)	Maturity total (°C h)	Cube strengths	
												24 h (N/mm²)	28 day (N/mm²)
10	4	25	12	30	2	20	70	360	50	110	600	4	45
10	6	35	10	40	3	25	135	400	100	135	770	7	43
10	8	50	8	55	4	30	240	440	170	100	950	10	42
20	4	40	12	45	3	25	120	540	80	110	850	10	42
20	6	50	10	55	4	30	210	550	170	100	1030	14	39
20	8	45	8	70	5	35	340	560	250	80	1230	18	36
30	4	55	12	60	4	30	170	720	180	100	1170	15	37
30	6	65	10	70	5	35	290	700	260	80	1330	20	34
30	8	80	8	90	6	40	440	720	390	80	1630	25	33
40	4	70	12	75	5	35	220	900	280	80	1480	20	35
40	6	80	10	90	6	40	360	900	390	60	1710	25	33
40	8	95	8	105	7	50	540	840	540	40	1960	10	15

of the circuit of the heating equipment and the process can be automated. As advised earlier, automated processes should be regularly audited and program selections recorded as well as recording any changes made together with the reasons.

Moisture gradient control is not such an easy matter. Cast-in induction type hygrometers can be used but they are very expensive and are unlikely to be extracted for re-use. This does not preclude one from using them as pilot test control items. However, generally speaking, close-packed thermocouples at these critical sections, mainly near surfaces, can give an indication that the concrete is drying out too quickly. As concrete dries out, its thermal conductivity will drop rapidly and increasing temperature gradients between thermocouples is a sign of this drying process proceeding too rapidly.

The temperature matched curing bath (TMCB) is a useful maturity indicator as thermocouples from the product can be used to control an identical cure for concrete cubes. These cubes can be demoulded and crushed before demoulding the concrete product as the cube results indicate that a design demoulding strength has been achieved.

The rebound hammer can also be used to indicate that a product is safe to be lifted and handled. The mimimum rebound number should ideally be obtained for the specific concrete in question using a works-generated calibration graph.

One precast concrete factory used ultrasonic testing for what was thought to be a strength indication. This relationship is somewhat tenuous and is not a recommended tool for this purpose. What the testing achieved was probably an indication of uniformity more than anything else.

Proof strength testing of the product is also a means of establishing acceptable maturity with ultimate load testing not being a necessity although, all too often, an attraction.

7.6 Indicative guidelines

It is stressed that the list that follows comprises guidelines and not specifics:

1 In steam curing chambers and 'cloches', maintain rh at 75–90 per cent.
2 Start with the mix as hot as possible.
3 Where hot mixing water is used, add the water before cement is introduced into the mixer and mix with the aggregates before adding the cement, continuing the mixing.
4 To heat a cold mixer, pour in about a litre of methylated spirits and set fire to it. Do not use petrol, diesel or other liquids as these will generally leave a carbon deposit. The Health and Safety Officer should be consulted before undertaking this practice.
5 Start the heating cycle as soon as possible after compaction.
6 Heat the concrete at up to 20K/hour for at least the first two hours.

7 Restrict temperature gradients within the range 50–150K/m depending upon the thickness of the section.
8 Restrict maximum temperature holding time to a minimum for the optimum cooling period required.
9 Do not let the concrete exceed 90°C.
10 Cool the concrete at a rate within the range 10–20K/h with a maximum gradient in the range 50–150K/m depending upon the geometry.
11 Use chemical self-hardening release agents preferably.
12 Surface retarders, when used, should be organic solvent-based.
13 Assess in a full-scale production unit.
14 Re-assess when materials and/or geometries change.

8 Hot and cold climates

8.1 General

Although the first edition of this book was intended for UK practice in precast concrete manufacture, it abstracted experience and showed illustrations from European and North American works and site visits considered relevant to UK work. Since 1982, the author has accumulated additional experience in precast concrete work in the Middle East and has applied this to a series of recommendations. These recommendations are based upon common sense and most of them will either appear to be immediately obvious while others will turn out to be so later in the section.

The sub-section on cold weather work is based upon a small amount of UK winter work in both precast and ready mix concrete and upon reported work from such areas as Siberia and Canada and various permafrost areas.

The UK precast industry has become more widespread over the past couple of decades and not only have companies opened up works in other countries but they are also exporting their products overseas. These export routes, surprisingly, have included air transport, although the products concerned were not huge in size. Thus, it is to be hoped that the discussion that follows will be found helpful to organisations either currently involved or thinking about such future activites.

Global warming is a current theme and continues to dominate the press and the political scenes and to affect the current way electricity is generated, albeit in small part only. In preparation for a global warming that might take place there is little that needs to be done over the next hundred years for a 5K climatic rise in average temperature but the predicted accompanying rise in sea levels could well lead to a demand, for example, for sea defence and vineyard support precast concrete units. The author's personal view of these predictions is that there is an indication as at 2006 that the climate is getting warmer but no more than that necessarily expected in the current inter-glacial ice age period. If such evidence is produced in incontrovertible form in the years to come, human emission of greenhouse gases is too convenient an item to blame. Industrial and domestic production of such gases 700 years ago could not be blamed for the warming occasioned then.

8.2 Hot climates

There is a tenuous relationship between accelerated curing and concrete work in hot climates and the aim of this sub-section is to promote control over situations where operators think that hot conditions help in the production but all too often matters can get out of control. The previous chapter dwelt at length on the chemical exotherm reaction of cementitious material hydration so, with the mix starting off at, say, 40°C, this reaction will be accelerated considerably.

Hot concrete problems in countries having, generally, more temperate climates can be tackled in a number of ways that are not necessarily available to hot climate countries or are available at considerably extra cost. For example, European and North American facilities are generally not too far away from coal-burning power stations making PFA available or too distant from steel-manufacturing factories where GGBS can be produced. Both these additives can be used for their hydration rate reduction properties. In hot climates there would seem to be three avenues available either singly or jointly:

1   controlling the material temperatures, material including the ingredients as well as moulds, reinforcement and all else;
2   local manufacture of low-heat cements;
3   working at the coolest time (including night).

Many areas of the Middle East have primary aluminium manufacturing facilities. The use of low emissivity materials such as aluminium foil and low emissivity coating such as aluminium paint needs to be explored as extensively as possible.

Middle East winter temperatures are not all that different from warm UK summer conditions and overheating precautions are generally not required. It is during the other seasons that problems arise with summer air temperatures reaching as high as 50°C. Aggregate storage should be kept under cover to minimise heat gain from both solar and wind sources but not under just ordinary cover. External storage stockpiles should be covered with heat-reflective sheeting such as aluminised blankets. Aluminium foil would be much more heat reflective but would be easily damaged and not as cost-effective. Internal storage should be in buildings with heat-reflective roofs and with façades made heat-reflective by aluminium or similar painting.

Lighting should be by low-heat emission fluorescent or similar, and windows either not incorporated or their areas kept to a minimum and solar reflective glazing used.

Cooling water pipes can be passed through the aggregate storage as an additional or alternative way of keeping the temperature down. A more scientific approach would be to use the latent heat of evaporation of water to cool aggregate stockpiles. The method would be to spray the stockpiles with as cool as possible water last thing in the evening and expose the aggregate to air during the night. Stockpiles should be temperature monitored to ensure that the aim is being achieved and that stockpiles are protected from warming winds as much as possible.

Provided that other properties of the fine aggregate used are within specified limits, natural sand is acceptable. The latent heat of evaporation of washed sand is a useful cooling aspect but the wash water should be fresh water or of an acceptably low salt level. Crushed rock aggregate, depending upon the precast process, will generally need the dust removed. This could be by washing but consideration could be given to cyclone blowing, keeping all materials nominally dry.

Where there is a choice of equal-performing aggregates from different sources and one of these sources is of a lighter colour than the others, the lightest colour one, if it is shown by temperature monitoring to have a significantly lower emissivity, should be selected.

Cement silos should be coated with heat-relective paint (as should ready mix concrete delivery trucks) and cement delivered in as un-fresh state as possible so that it has had maximum time to cool down following its grinding. Consideration could also be given to the manufacture of hollow-wall silos so that cooling water can be circulated in the jacket of the silo.

Cements tend to be ground finely to produce rapid-hardening behaviour and a trend to coarser ground cement production in hot weather could well be found to be a 'low heat' advantage. In some forms of machine-intensive precast concrete production, having a cement too fine could well be found to be a disadvantage for reasons other than its exotherm properties. Should a specific low heat cement be available, then this could well be an advantageous ingredient to purchase.

Moulds left in the sun will obviously get hot and transfer this heat to the concrete placed therein; therefore moulds should be kept under similar cover to aggregate storage and maintained so during the setting and early hardening times. Steel and GRP moulds, having relatively high thermal conductivities and specific heats compared to timber or timber composite moulds, are particularly prone to get hot quickly. Steel moulds are not easy to paint with aluminium paint but such treatment can help in resisting heat solar heat gain. Aluminium powder can be incorporated into the finishing polyester coat of a GRP mould, acting as a permanent heat-reflective layer. The exteriors of these moulds would need to be kept clean to maintain this property.

Mixing water should be kept as cold as possible. Having the highest specific heat of any of the concrete mix ingredients, it is, as with accelerated curing, generally the most viable material of the mix to heat as well as to cool. Iced water, supercooled pressurised water cooling tubes or crushed ice are at least three of the ways of cooling the mix water, ensuring that crushed ice entering the mix thaws out before the concrete is used in the production.

Reinforcement, prestressing, cast-in hardware should be subject to similar heat protection from the elements. Where the design permits, consideration can be given to the use of hollow reinforcement through which cooling water can be passed. Cooling tubes not intended to act as reinforcement can be placed in the mould, bearing in mind that their design should not detrimentally affect the mix design (maximum aggregate size and workability) requirements.

Production during the lowest hot climate day/night temperature has its attractions as there is no solar heat gain and the only precaution that might be necessary is against hot winds. It therefore behoves one to plan on night-time production to include the whole of the demoulding, preparation and casting cycle. This means a management and operative philosophy geared to a changed approach to production between winter and the hotter seasons.

Curing, both from the temperature and moisture considerations, is extremely important. Solar radiation and/or drying winds should be prevented from impinging upon exposed fresh or hardening concrete faces. Protection with weighted-down foil-faced expanded polystyrene covers is one of the ways of inhibiting too high temperature or moisture gradients at these faces. It has to be borne in mind that such covers need to be removed as soon as possible to allow cooling without too fast a moisture loss. This can be achieved by polythene sheet or similar covering.

As with accelerated curing, the concrete should be temperature-monitored and, where possible, moisture-monitored to ensure that the product is kept well within the critical limits that could cause cracking or other distress.

8.3 Cold climates

This sub-section is much shorter than the previous one because of both lack of the author's experience as well as the low likelihood of a UK precaster carrying out winter production in areas such as Alaska or Siberia.

Washed aggregates here will be ice-containing. There was a process in Russia (no knowledge of its current use) of mixing and casting ice-containing concrete and placing it into moulds and waiting for warmer weather for it to thaw and be used. There is a great danger of producing voids in the product unless re-compaction is applied and this invokes a timing problem of picking the right thaw condition before applying this re-compaction.

It would appear that the general problem is singular and that is starting with the concrete as warm as possible above 0°C and keeping it so. Thus, hot oil or hot air pipes circulating through the stockpiles, using timber or sandwich insulated steel or GRP moulds coupled with rapid-hardening cement as fresh and hot as possible are some of the precautions that would need to be taken. Premises should be well-heated and insulated and glazing areas kept minimal or with no glazing at all.

Mix water should be heated and added to the mix and mixed before the cement is added. Depending upon conditions this water could well be near its boiling point. It needs to be remembered that as plant heats up to a static temperature, following the first one or two mixes, that this water temperature needs to be reduced.

Although production could well be in permafrost areas, a geological survey could establish where the geothermal rock strata depth begins and a well be dug to use this heat in the production as well as to heat the factory. Reference was made in the previous chapter to the use of methylated spirits to heat up plant and equipment and similar consideration should be given to cold climate work.

With hot climate work reference was made to a nightime/daytime change in production approach between seasons and could well be that in cold climates it might be most viable to close down for winter production completely and only begin in spring or summer with ambient material and air temperatures above freezing.

Several areas of the world that have hot daytime temperatures have very cold nights. It is possible that the psychrometric behaviour of water/air mixtures at differing temperatures can be put to good effect.

Although cold air cannot hold as much moisture as warm or hot air, there is always a difference in the vapour pressures between different combinations of air and water. There is a very large temperature inertia effect in the ground and atmospheric conditions do not affect the ground to a very large depth. It is estimated that the temperature at depth in the ground in tropical/semi-tropical/desert regions is in the static range of 10–15°C and there is always moisture present at these depths. Therefore, with cooling nightime temperatures down to, say, 5°C, the moisture in the ground is in an environment at a higher vapour pressure than the atmosphere and will tend to migrate towards the surface.

If this moisture can be retained by covering the ground, the vapour pressure will tend towards saturation and condensation will occur. It is possible to dig pits into the ground and line them with vapor-permeable membranes as used in the UK as roof tile underlay. One could then place the day's production in the pit and cover it to prevent excess evaporation.

9 Properties and performance

9.1 General

This chapter and the one on accelerated curing are the only ones that use the same titles as the first edition of this book. Both these chapters were considered too singular in nature to be integrated into other sections as has been the approach for the rest of this book. To place properties and performance as a single chapter content was found to be a daunting task in the first edition and no less currently. Although some of the same sub-titles have been employed for the aim of improving readability some of the items have been integrated into more compact sub-sections.

Tunnel vision in construction practices has been a failing in many instances and no less so in the precast concrete industry. It has to be borne in mind that the finished item, be it anything from a 1000T bridge support unit to a 2kg garden gnome comprises, a multi-variant application of materials, plant and labour resulting in a multi-property end item. The failing in the attempt to achieve good quality is often due to the concentration on one property, or perhaps two properties, for all that one thinks is necessary and ignoring other significant and relevant properties that govern performance.

Several other practices occur giving rise to self-generated problems and examples of these are given in the list below. It will be seen, hopefully with later reading, that the whole aim of this chapter is to promote a balanced view on all matters relating to properties and performance and encourage the parties involved in precast projects to liaise with each other to ensure that sensible and relevant practices are followed resulting in overall mutual benefits:

1   Placing over-reliance on cube test data without studying how the cubes were made, cured and tested, including the status of the test machine and the operatives involved.
2   Accepting a test requirement that has no relationship to the product or its performance. For example, asking for a freeze/thaw requirement for cast stone is superfluous as the product has never been reported to have suffered from such damage.
3   Accepting a requirement for a treatment such as silane application when the concrete is too impervious to absorb the chemical.
4   Accepting recipe and performance requirements jointly. The performance with its related properties is what should be specified. It is then up to the expertise of the precaster to produce that which is required.

This list could be extended significantly and it is left to the precaster to take necessary steps as and when required. At the expense of causing complications, these steps could well include drawing to the other parties' attention the possible omission of a test requirement considered to be relevant to performance but omitted in the specification documents.

At this relevant juncture a favourite hobby horse of the author can be taken from its stable and exercised. The construction industry is a potential gold mine for consultants, testing authorities and the legal profession as errors continue to be made and the same errors from years ago repeated. It is opined that the reason for this is politically originated as the construction industry has been used as the most convenient financial regulator in any industry. Of the estimated more than 30b/annum turnover in the concrete world, possibly about a half originates directly and indirectly from government direction. As a short-term adjustment exercise it is easy for the government to increase or decrease the monies available by a billion pounds or more. This has happened for decades and is not the remit of any particular political party. The results of this policy at least the following:

(a) Industry cannot plan for levels of ongoing workload.
(b) Training in employment of the professionals cannot be a long-term feasible matter as used to be the custom.
(c) For all levels of potential employees the bleakness of future employment is a deterrent.
(d) Wages and salaries reflect these matters by being unattractive.
(e) Permanent staff employment is no longer so widespread and much of the work undertaken is by sub-contract or agency or similar and control is not necessarily as good or easy as it used to be.

Before continuing, it is stressed that although each following subsection has its own title, the dividing lines between them are not necessarily sharp and distinct. Having come up with these five criticisms (there are probably a lot more), readers might well wonder why one should bother with properties and performance. The answer lies in the author's optimism in that things will get better with the whole of the industry taking a more pro-active role in determining its own future.

The main message that this chapter hopes to reinforce is that the precast concrete industry can produce high quality products and, just as importantly, can prove it.

9.2 Mechanical

Product strength (proof test or better known as 'realcrete') or cube strength (type test or better known as 'labcrete') attract the most attention and specifications relate either to proof or type requirement. The precast industry has an advantage over the in situ section in that a 'green' or early strength is necessary to run a viable production. This results, more often than not, in a surplus to the 7- or 28-day specified strengths. The cost of testing is an important factor for relatively large products where proof testing without destruction should be the norm. If proof testing were to be taken to destruction, there would likely be a cost factor involved. Should such a requirement apply, the cost of the destroyed unit could well be an item in the bill of quantities charged to the client. For products such as blocks or roofing tiles, testing several of them would not necessarily show as an extra above acceptable breakages.

The Standards that cover precast products are generally proof testing in nature. For large units such as beams, columns, etc., cube type testing is often undertaken. As emphasised in Chapter 7 on accelerated curing, type testing of cubes for the indication of the strength of these large units should always be with the cubes cured under as near as possible in identical conditions to the product. It is also a good idea to assume that cubes are generally easier to make than the actual product and some allowance be made for a strength difference even with virtually identical curing conditions.

Apart from most concrete blocks being tested in compression, the vast majority of specified proof testing is in flexure which is of benefit in two respects. First, it relates to the early handling properties of the product, a major requirement. Second, it shows the precaster that there is an early flexural strength bonus in that the later nominally fully-hardened compressive/flexural strength ratio of 10/1 can be nearer a figure of the order of 7/1 at 16–24 hours old.

Where cube or cylinder type testing data is being assessed, the sample is weighed and 'density' figures are given. For precast concrete products, density data is of little or no value for the simple reason that the sample dimensions can vary, especially on the height to the floated face. In addition, the sample is weighed wet and this relates to both its water absorption as well as to how thoroughly the operative wiped it with a damp rag before weighing and crushing.

Cores can be taken from a product and tested but the data is not easy to interpret as may be seen in the Concrete Society's comprehensive report on in situ concrete project on cube/core relationships.[52] The author's opinion is that coring precast products should be discouraged except, possibly, where one is looking for specific effects such as air entrainment, bonding to reinforcement, petrological properties and similar.

Products that often have complex geometries such as cast stone will generally need to be type tested as only a few products are shaped as rectangular parallelapipeds, thus making themselves amenable to compressive or flexural testing. It is probable that due to the differences in compaction, more differential allowance needs to be made between a cube test result and the product strength. As moist mix design cast stone products are generally mortar mixes, a well-compacted product will typically have a density in the range 2000–2200 kg/m³. It has not been unknown for some products to have questionable compaction with densities in the range 1700–1900kg/m³ with subsequent reduced strength and durability.

The mention of the word 'durability' is a reminder to conclude this subsection with a warning that strength, proof or type, has little or no relationship to the vast majority of durability hazards. First, one needs to define durability as to 'durable to what?'. Second, there is the necessity to examine all the hazards for a product's specific application. It may well be found out that the higher a product's strength, the worse its performance. Examples of this are reduced ultimate strain capacity due to lower elasticity, brittleness in handling resulting in damaged arrisses and so on. Most product and system geometries benefit by keeping targets within a range rather than looking for maxima or minima limits.

9.3 Physical

This sub-section discusses the physics of concrete. Physics, like chemistry in the next sub-section, is a rather demeaned subject as engineering has been dominant and only over the past decade or so have people begun to realise that physics and chemistry are not engineering (civil and/or mechanical) subsubjects. There are several subjects in physics relevant to the performance of both the product *per se* and its behaviour in the works:

1   pore structure and permeability;
2   hygrothermal, moisture transfer and condensation;
3   thermal conductivity and inertia;
4   sound insulation and absorption;
5   fire resistance and reaction.

9.3.1 Pore structure and permeability

The majority of durability hazards often relate to pore structure which should not be confused with porosity. It is possible to have an impermeable porous material if the pores are discrete and not interconnected by capillaries or other paths for fluids to pass. Foamed glass is an example of an impermeable porous material. Concrete falls into the permeable porous class which can result in a risk of carbonation, sulfate, chloride, freeze/thaw and organic acid attack as would be encountered in many agricultural applications as well as used in moorland water situations.

The author's book[53] put into context the ignorance of facts in the panic that related to the use of high alumina cement, now called aluminous cement. A few words about the physical listed hazards from the previous paragraph might not go amiss. The incidence of damage due to any of these physical effects in the UK is minute. The unfortunate thing is that when there is a reported problem with concrete, it receives undue attention both in the press and in the services offered by organisations in dealing with the alleged problems by consultancy, testing, running seminars and conferences, and so on. At the time of going to press of this second edition, problems have been reported with the foundation concrete of a stadium construction. Problems can arise from either materials and/or design and/or workmanship with there being not necessarily a sharp division between the last two. Analyses of 'trouble-shooting' items over nearly two decades in the era of Laing R&D showed that materials reasons constituted less than 15 per cent of the total. The author, with a reasonable degree of confidence, predicts that the stadium problems were due to design and/or workmanship.

The previous edition went to great lengths to describe the mathematical bases for the various assessments of permeability and, not being rocket science, it is not proposed to repeat these in details but to highlight what are considered to be the critical points. In-depth reading of all these tests is discussed in the Concrete Society's Technical Report,[54] which, in 2007, is due for updating.

Initial Surface Absorption Test (ISAT)

The old BS1881 test is referred to in the Concrete Society's report,[55] studied and reported at length by the author[56] and is called up in the Specification for cast stone.[57] It consists of sticking a knife-edge cap or clamping a gasketted cap to the prepared surface and subjecting the volume therein to a 200mm water pressure and measuring and recording the ingress speed at specified intervals from the start of the test. This pressure simulates a driving wind and rain at nearly 100 km/h and therefore corresponds to the worst of rain-driven conditions likely to be experienced in the UK. The test has been contract specified[58] for both precast and in situ concrete and the respective relevances of its value can be seen on the exposed aggregate precast cladding on St. Katherines-by-the-Tower Hotel by Tower Bridge and the large plain flanking walls at the Carlsberg Brewery in Northampton. As one of the cast stone specified tests, its value can be seen on many contracts with special reference to the relatively new 'cottages' in the Inner Circle of Regents Park in London. An older building, circa 1970, is Holy Trinity Church in Hounslow, West London. This church gained a special dispensation for its altar which is also made of cast stone. Small samples of cast stone containing increasing concentration of water repellent were placed on the ground at Wexham Springs to weather for over a decade. With aluminium stearate integral water repellent concentrations from zero through to 2 per cent m/m cement Figure 9.1 shows algae growth and dirt decreasing with increasing water repellent concentration over this period.

For longevity of performance of cast stone which, had it been permeability tested at the time, the reader's attention is drawn to the 1875 Loanhead Church building on the South-East side of Edinburgh. The building is not thought to be currently used as a church but has an amusing omission. There is no clock face on the side facing the fields allegedly to inhibit 'clock watching' by farm labourers. Illustrations are shown in Figures 9.2, 9.3 and 9.4.

Applications of the ISAT to precast and in situ concretes are shown respectively in Figures 9.5 and 9.6. The first illustration is on cast stone coping units and the second on an in situ concrete footbridge beam.

Figure 9.1 Effects of 10-year weathering on moist mix design cast stone with increasing levels of integral water repellent.

Figure 9.2 Example of Loanhead Church, near Edinburgh, built in 1875 with cast stone units.

The absorption test

This test is normally applied to test cubes but can equally be applied to whole units or specimens cut from whole units. It consists of oven-drying, cooling, weighing and immersion in room temperature water for, typically, 30 minutes. It is not a very meaningful test because of the following:

1   The water accesses faces including those not subject to weathering (e.g. cut or sawn or cored).
2   The accuracy of testing is to within 0.1 per cent of the dry weight. For a typical 30-minute range of data from 1.0–3.0 per cent, this does not offer much of a 'good', 'fair' or 'bad' assessment.
3   The test is sensitive, obviously, to how much excess water is wiped off before weighing.

Figure 9.3 Example of Loanhead Church, near Edinburgh, built in 1875 with cast stone units.

Figure 9.4 Main door to church shown in Figure 9.3.

Figure 9.5 ISAT on cast stone coping units.

4   Cut or cored samples could well suffer incipient damage in their preparation, giving rise to a higher than true figure.

Capillary absorption

This test is specified in the masonry unit standard[59] as well as an alternative to the ISAT in the British Standard for cast stone.[60] It consists of placing the prepared product face down in a few millimetres of water and measuring the amount of water picked up by capillary attraction. It was introduced into the previous edition of BS1217 on cast stone as clients wanted some products to exhibit weathering more quickly than cast stone units complying with the more stringent ISAT specification. It has a value in cast stone in that it shows which products contain a water repellent and which do not but, being at a very low pressure head (a few millimetres), it does not tell one how effective the water repellent admixture is in resisting weather. A glass thistle funnel with a welded glass tube stuck to the surface with greased 'Plasticene' and filled in 10mm steps with water will show, at breakdown, how water repellent the product is.

High pressure water testing

Although this test probably has more relationship to in situ concrete than precast, it can be used for precast products which are designed to be used under

Figure 9.6 ISAT on side of in situ concrete footbridge beam.

high hydraulic pressure. The test has to be carried out on laboratory-made samples or cores drilled from the product, all tests being under laboratory conditions. The test is relatively expensive compared to the above-described ones, but does give an indication of the strength of capillary and pore walls.

9.3.2 Hygrothermal, moisture transfer and condensation

Although this particular subject relates to the construction rather than the product, site performance relates to the items making up that construction. The interest is how the construction performs under specified conditions of temperature, relative humidity and material properties. Specific interest would apply to constructions such as beam and block flooring, walling and flat roof constructions. A computer modelling approach is required which can deal either in 2D or 3D modes. The factors to be examined assume that the construction meets the Regulation requirements and shows:

1. temperatures at all points with emphasis on the coldest; internal face point(s);
2. critical temperature factors and their relationships to both mould growth and condensation risks;
3. compliance with the overall thermal insulation needs.

9.3.3 *Thermal conductivity and thermal inertia*

Tables of concrete thermal conductivities are given in the Standard[61] but these should not be used as they stand. They are for nominally oven-dry concrete, whereas concrete, on site, precast or in situ, is usually classified as protected (indoors or sheltered outdoors) or exposed (outdoors and subject to the weather). Values of conductivity are necessary in any modelling or calculation work and manufacturers should always record the provenance of any figures used.

Concrete has the general advantage of having a high thermal inertia in that a typical density concrete of about 2300 kg/m^3 takes a lot of time to heat up and a lot of time to cool down. A cavity wall construction with its leaves reversed might well have identical U values (thermal transmittance) but the wall with the higher density internal leaf will give more comfort because temperature changes will not be sudden and thermostat switches will not have so much work to do. This would probably lead to the switches having a longer lifetime. In comparing a 'reversed' cavity wall construction, the lightweight outer leaf would have a higher moisture content than if it had been used as an inner leaf. The reverse would obtain for the denser inner leaf. However, where the benefit lies is that the damp outer leaf will have a lower thermal conductivity than the nominally dry inner leaf. Thus, the overall thermal transmittance of the reversal would be lower than for a conventional brick/block construction.

9.3.4 *Sound insulation and absorption*

Three forms of sound insulation are of relevance to precasters:

1. air-borne;
2. impact;
3. reverberation.

Air-borne sound

Precast concrete has an air-borne sound resistance characteristic that relates, approximately, to the weight per unit area. In effect, a wall or floor of density x and thickness y has, approximately, the same air-borne sound insulation as a wall or floor of density 0.5x and thickness 2y. In designing constructions aimed at compliance with Regulations or Specifications, it is unlikely that the concrete by itself would form the sound barrier. Linings, secondary leaves, cavities and other details of construction come into play and testing on site will normally be the most logical approach and is often applied. Current Regulations have a dual approach of a recipe in the design known as Robust Construction Details for Sound Insulation or a site test is invoked. Such testing will bring into consideration not only details of the design but will also pick out workmanship aspects, especially that of flanking transmission.

In housing construction, the interest lies mainly in party wall performance and sound transmission internally within a dwelling from one part of the dwelling to another. Concrete blockwork used in a party wall will be detailed with linings and particular cavity design so the concrete *per se* is contributory to insulation.

Impact sound insulation

The density property here is virtually the opposite of air-borne sound transmission with the transmitted sound arising from an article/body striking the material, The noise generated by the impact travels through the material and is emitted as air-borne sound at the other side. Therefore, the lower the concrete density, the more able it is to absorb impact energy and cause less sound transmission. This is why a beam and block floor construction is effective in inhibiting both air-borne and impact sound transmission due, respectively, to the mixture of high density beams with relatively low density blocks. It also needs to be remembered that good impact sound absorbers are also good sound emitters. A poor design would be to have something like an EPS layer on the soffite of a solid floor. A suspended ceiling of the same material would behave completely differently (ignoring the fire regulation requirement).

Reverberation

The mechanism here is that of surface texture with high reverberation levels being associated with conversation inhibition in rooms with large gatherings and effects such as unpleasant echoes in concert halls, cinemas, churches and similar buildings. Texturing the surface is one method of dealing with concrete with exposed aggregate internal surfaces being successfully deployed as at the Barbican, London, concert halls and cinemas.

9.3.5 Fire resistance and reaction

Fire resistance

Concrete is a virtually fire-proof material and failure in test or in practice depends upon the time where the first of the following observations occur:

1 structural integrity is lost;
2 passage of flame occurs;
3 the 'cold' side reaches a surface temperature that can ignite neighbouring materials.

Structural integrity depends upon thickness and reinforcement/prestressing design. As a rule of thumb, conventional dense concrete resists fire for 1 hour for each 25mm of thickness. Lightweight aggregate and aerated concretes, having lower thermal conductivities, would fare better. Reinforcement or prestressing will be detrimentally affected if it is too close to the fire surface and secondary reinforcement to hold the protective layers of concrete in place may be required.

Laboratory testing is normally carried out under design loads. In practice, fires in concrete constructions do not generally impede fire crews from entering the building to fight the fire as reinforced concrete will tend to have a slow collapse and hogging of prestressed units is an early evacuation warning. Thus firefighters generally have adequate warning as to when it is no longer safe to stay inside a building.

Fire reaction

Standard tests for the general run of construction products are designed to pinpoint emission of noxious and/or harmful gases generated by the fire. Concrete is generally innocent of all such materials in their ingredients with the exception, possibly, of sandwich panels containing expanded polystyrene.

9.4 Chemical

The hydration of cementitious products is a chemical reaction and most of the chemicals involved hydrate to form calcium silicate hydrates and other hydrates. Nevertheless, two chemicals are involved which can go into solution and, fortuitously, contribute to the alkaline nature of the matrix and the ensuing passivity of steel and its consequent resistance to corrosion.

The first of these chemicals is the least harmful and is the release of hydrated lime in the hydration process. Lime is only slightly soluble in water and what goes into solution assists in keeping the pH value of the matrix in the highly alkaline 12–13 range. Lime solution in contact with the skin can cause irritation and, in some individuals, lead to further complaints. Due respect should be given when handling fresh concrete with eye washing applied when concrete is splashed in one's face. Whether in a precast works or on site, eye washing stations should be situated at critical points and all operatives trained in their use.

The high risk chemical group is the alkali oxides of sodium and potassium given the symbol R_2O. Although they constitute a typical 2–3 per cent mass content in Portland cement, they are both very soluble in water. The subject of cement burns was treated at length in the book on concrete materials[62] and was shown to be the major reason for removing fresh cement-based materials from the skin and clothing. The burns are necrotic in nature and the nerves in the skin area would appear to be anaesthetised by the alkali (note: the word 'alkaline' is not used) and cannot feel the burning away of the skin and flesh.

Dangerous leaching of these alkalis can occur and has been observed in hardened concrete on site. The case in question was of year-old indoor prestressed hollow core floor planks which were observed to be slightly sagged when loaded with internal walls and other floor loadings. The inverse soffite was noticed to be damp and an operative drilled into it (fortunately with a low voltage drill) to drain the water but it was not water but a concentrated caustic soda/potash solution which discharged and burnt his hands quite severely. This leachate had been dissolving alkalis from the surrounding concrete for a considerable time.

There is another chemical ingredient in cement, namely hexavalent chromium where its soluble salts are harmful to most forms of animal and plant/marine life. This was also discussed in the materials book.[63] Cement manufacturers generally deal with this chemical by adding extra iron-based chemicals to the cementing ingredients or employing other means of complexing chromium into an insoluble and/or relatively harmless form.

What alklines, alkalis, mould release agents and construction chemicals in general mean is that both personal and family allergic histories of all potential and existing operatives be examined and recorded and tasks be allotted to suit that person's health and safety requirements.

9.5 Aesthetics

9.5.1 General

What concrete looks like is within the province of the design and manufacturing teams and the bad name that concrete often gets is more often than not due to 'spoiling the ship for a ha'porth of tar'. The subject of aesthetics generally relates to the external façade appearance but not necessarily to the exclusion of internal façades. Generally, on cost breakdown, the contribution to the total construction cost of façade elements is small compared to the total building/construction frame cost which, in turn, is generally small compared to the fitting out costs. Therefore, it logically behoves the construction team to plan ahead in this respect and design for appearance.

There is nothing particularly amiss with a façade becoming and staying dirty if that was the architect's intention and there can be an aesthetic attraction in such an appearance. Where possible mistakes can be made in how dirt affects buildings is in not designing them for the dirt to channel itself in purpose-made areas and/or not taking account of the prevailing wind and rain conditions of a façade. The elevation facing the prevailing wind and rain will have a different weathering behaviour to the more protected elevations. These latter elevations will generally hold dirt and staining for longer periods than the more heavily exposed façades.

This is where geometry comes to the aid of the designer. A vertically-troughed unit tends to have its returns in shadow and since this is where rain will run and dirt collect, such a troughed design assists greatly in the weathering appearance. Similarly an exposed aggregate finish will tend to promote dirt collection in the macro and micro spaces between the primary aggregate. Probably the worst architectural appearance one could design without taking appropriate mix design precautions is that of a smooth so-called fair-faced finish. This effect can be easily exemplified in two well-known structures, both of which are several decades old. These are two London entertainment centres illustrating the opposite ends of this visual spectrum, the South Bank Festival Halls and the Barbican Centre.

A degree of hyprocrisy could be said to prevail in the case of comparing cast smooth-faced Portland Stone finish with natural Whitbed or Cap Portland stone where, for the cast variety, a boring sameness may be specified whereas, for the natural product, a stone-to-stone variation would be expected.

9.5.2 Specifics

It was decided to set out this sub-section in a series of short sharp points with several illustrations, this being a more readable form of approach. For in-depth valuable and detailed discussion, attention is directed to the US Prestressed Concrete Institute publication on architectural finishes.[64]

Specific problems that affect the aesthetics are:

1 Staining due to lime bloom, often called efflorescence, is always a latent hazard as there is generally an excess of lime available from the cementitious components. The more permeable the product, the more this is likely to occur and the darker the product, as shown in Figure 9.7, the more is the comparison effect. It is unwise to make a highly permeable aesthetic product (as moist mix design cast stone) without incorporating a water repellent admixture in the mix.

There are different schools of thought on the use of hydrochloric acid for the removal, the author supporting the 'don't use' one and allowing

Figure 9.7 Effect of lime bloom on dark-coloured products.

weather to take its course. Any residual chloride from the acid left on the product will render the surface hydrophilic and cause it to promote more lime bloom.

2   Staining due to hydration staining was discussed in the author's later book[65] and it arises from using glossy surfaced moulds. Such moulds are best made slightly matt before use, often achievable by running 2–3 dummy casts before full production. The staining is due to a physical and not a chemical process and the effect can penetrate 5–15mm deep into the surface and can only be removed physically by cutting out the defect. Products are difficult to demould as the defect is thought to be micro and not macro in nature with adherence to the mould being due to van der Waal atomic distance attraction. This will occur no matter what mould release agent is used. Where hydration shrinkage results in the concrete shrinking away from the mould faces, air containing carbon dioxide will ingress and carbonate the surface, giving rise to the typical grey concrete colour. The colour of hydration staining is that of a dark slate grey to approaching black in appearance.

To simulate hydration staining, place and compact concrete into two separate glass jars and seal one with an airtight lid. The sealed jar will exhibit hydration staining round most of its visible surface whereas the unsealed concrete will show the customary grey colour.

If one wished a more scientific approach, clean and polish two pieces of copper to a mirror-smooth finish in a nitrogen box and hold these faces together for a second or two. The bond obtained will generally be stronger than a brazed or soldered joint. Another example is that if 35mm slides are being mounted. It will be found that the glass squares are almost molecularly smooth and have to be slid off to part them. It is virtually impossible to prise them apart.

3   Staining of relatively mature concrete due to organic sources such as mould release agents, leaves, timber, are best left alone as the free lime in the product will generally bleach them out due to a saponification process.

4   Staining due to site detritus, as shown in Figure 9.8, for badly-stored cast stone sills, is likely to be a life-long disfigurement. Any removal trials should be assessed on small low visibility areas before embarking on a total surface approach. A main site risk is that of bitumen which is used for rear surface treatment of the external leaf of cavity construction and which, for some reason seems to have a magnetic-like attraction towards visual surfaces.

5   Product-to-product surface appearance variations will nearly always occur, even with the best of controls. A full size mock-up including all the variables that are to be included in full production should be offered for inspection. Figure 9.9 illustrates such a mock-up where it is just possible to discern changes in shade. Figure 9.10 shows a mixture of 3-month-old exposed aggregate and fair-faced units on a building illustrating what is possibly nearer the norm for the head units. Figure 9.11

Figure 9.8 Architectural unit badly stored on site.

Figure 9.9 Mock-up of architectural panels.

shows the same variation for moist mix design cast stone ashlar units on a Palladian elevation.

It has been found best not to attempt any form of surface treatment aimed at uniformity but to leave natural weathering to take its course, accepting that protected faces will take longer to achieve this sameness. Exposed aggregate surfaces achieved by washing, surface retarders or, as shown in Figure 9.12, by grit-blasting or other means tend to produce less surface-to-surface variations than for fair-faced units. However, care needs to be exercised to ensure that such processes are not used with poorly made products as shown in Figure 9.13 or taken too deeply as illustrated in Figure 9.14 where there is a risk of long-term matrix weathering resulting in aggregate dislodgement.

It is recommended that specifiers asking for plain fair-faced units should be advised that in-unit and between-unit variations in colour and texture will occur, even with the best control that can be exercised. If such a requirement remains necessary, then at least two full-sized mock-ups as illustrated in Figure 9.9 should be produced, one for the building site and one for the factory. The care exercised in these preparations needs to be typical of the general controls exercised in production. There is little purpose in producing a special sample that would be very difficult to match. In the same vein, precasters who use small samples (e.g. 100mm × 100 mm × 25mm) in their publicity should advise potential and existing clients that these samples are indicative and not necessarily a close resemblance.

Figure 9.10 Three-month-old façade of fair-faced and exposed aggregate units.

Another point that cannot be stressed too strongly is that variations observed in the early ages on site will generally weather down to a sameness over time. Depending on the building elevation, this weathering process can take from a few months to a few years. As remarked earlier, surface treatments are not recommended as all too often the apparent initial improvement is lost to later differential weathering.

6   Staining due to the application of too much release agent is shown in Figure 9.15. Although this is a non-visual product it would be thought that appearance does not matter. For visual concrete Figure 9.15 emphasises the need to keep release agent use to minimum coverage.

7   Aggregate transparency, as shown in Figure 9.16, is a problem with fair-faced concrete and is associated with the use of gap-graded rather than continuously graded aggregate mixes. This effect is especially noticeable with spar and light-coloured granite aggregates. Gap-graded mixes should only be used for visual exposed aggregate units.

Figure 9.11 Few-months-old Palladian style façade in cast stone.

8 Pigeon droppings cause severe disfigurement and horizontal surfaces can be wire-protected as shown in Figure 9.17. In other cases it is best to avoid using fascia units where there are visual areas of concrete where birds can stand.

9 Iron pyrites impurities, often encountered in flint and limestone gravel sources, become oxidised by the weather and cause staining as shown in Figure 9.18. Chemical treatment more often than not make matters worse and it is best to deal with these areas by physical removal followed by making good. Such remedial work should be undertaken at as young an age as possible so as to minimise visual comparisons. Ideal architectural matching in these repair cases is virtually impossible.

10 Crazing, as shown in Figure 9.19, scaled from the 50mm diameter lens cap that can be seen, is a disfigurement associated with light-coloured plain concrete and cast stone than with ordinary concrete. The photo-

Figure 9.12 Grit-blasting process to produce exposed aggregate finish.

Figure 9.13 Variable compaction in an exposed aggregate unit.

Figure 9.14 Over-exposure of aggregate in a visual unit.

Figure 9.15 Mould release agent staining.

Figure 9.16 Aggregate transparency in a visual unit.

graph shows a rather old (estimated 20 years) piece of cast stone where autogenous healing due to calcite formation in the crazing is taking place. Crazing is an aesthetic defect and never proven structurally nor to be a frost hazard. Its likelihood can be minimised and reference is made to Chapter 5 with its recommendations. The crazing 'polygons', scaled from the 50mm diameter lens cap shown in Figure 9.19 can be seen to be about 40–60mm across. Their surface crack widths always appear wider than they actually are due to the capillary attraction of moisture pulling the rain water in and leaving the dirt on the surface. Crack surface widths for typical crazing are about 0.2mm but the dirt makes them look nearer to 1mm wide.

11 Structural cracking is also an aesthetic defect as well as a structural/ corrosion matter. The defect is illustrated in Figure 9.20 and was considered due to exceeding the specified[66] slenderness ratio for a cast stone lintel. The tendency to manufacture long thin units in reinforced or unreinforced concrete should be resisted. Only prestressed or fibrous concretes can tolerate such demanding geometries. The transverse flexural dead load applied to these slender units when handling or storing can be temporarily doubled or trebled when lifting by personnel or mechanical means causes additional stress due to the acceleration-induced force.

Figure 9.17 Bird-repellent wiring on coping units.

9.6 Handling

The questionable handling shown in Figure 9.21 not only risks arris damage but can cause staining as lifting hooks and chains can be left on visual surfaces outdoors. Comments about health and safety are superfluous, at least the operative is wearing a safety helmet and gloves. Planning of lifting and handling both in the works and on site is necessary. Figure 9.22 illustrates a bespoke scissor grip used for picking up a batch of paving stones.

Handling precast products with minimum damage (or none, preferably) is generally a matter of common sense and two main aspects need to be borne in mind when dealing with young-age products:

1   Steel is harder than concrete.
2   Arrisses are easily damaged.

Figure 9.18 Iron pyrites staining.

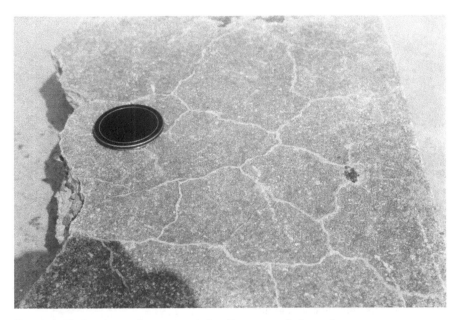

Figure 9.19 Crazing and autogenous healing on aged piece of cast stone.

Figure 9.20 Structural crack in cast stone with high slenderness ratio.

Figure 9.21 Example of poor handling.

Figure 9.22 Bespoke device for lifting paving slabs.

9.7 Performance

All too often performance is denigrated by the 'latest scare' which seems to put the industry onto its back foot and makes it act defensively. Performance, in all its structural, physical and aesthetic forms, is what sells precast concrete. The single point of view of what is delivered and paid for on site or in a builder's merchant being sufficient is wrong. The way a product performs is also a major function of design and construction. The precaster often has little or no say in what occurs after delivery and there is a latent danger of post-delivery involvement being misinterpreted as responsibility for faults that might occur that have, generally, nothing to do with the precaster. This does not necessarily preclude the producer from preparing and submitting a 'recommendation for use' document.

What performance requires is a team approach to construction where the precaster is involved from the design to completion stages rather than just attending the occasional site meeting where the subject is alleged trouble with the precast units. The previous sub-section listed a significant number of performance defects but, as stated earlier, the concrete industry with its precast component has been subject to an undue amount of adverse publicity. This is unjustified. An outstanding example of this was aluminous cement discussed at length in the book on materials.[67]

Suffice it to say that concrete in general and precast concrete in particular enjoy a good performance record and are predicted to carry on doing so with current practice being continued and supported by the parties associated with the industry. When it comes to performance, the attention of clients should be drawn to the track records of existing structures with an emphasis on those of the longest standing. The industry should respond vigorously to the poor publicity that the small percentage of defects causes by emphasising what such examples mean in the context of their rarity and guarding as far as possible against both repetition as well as rogue producers.

Performance assurance can be adjudged and proven in one or more of the following features.

9.7.1 Strength

Material strength, as distinct from design and/or workmanship problems, is rarely a problem with precast concrete units simply because the units have to be handled and stacked at a very early age. In effect, if an equivalent cube strength of the order of 10–20 MPa is not achieved at 24 hours old, a precaster could face an unacceptable reject rate due to breakages. Although strength and maturity are closely related, it has to be appreciated that hydration shrinkage continues long after adequate strength has been achieved. Thus, a unit strong enough to be built into a construction at, say, 2 weeks old, will still shrink away from its surroundings up to at least 28 days old.

Where strength is possibly of interest is in arris property as damage can occur to these corners quite easily. This means that great care needs to be exercised during handling and transport mainly because arrisses are under the worst of curing conditions.

9.7.2 Physical

As with maturity properties in general, improving properties in frost resistance, attrition, fire resistance and similar are slow to develop. Continuing hydration causes the pore structure to improve (lower permeability) and so immature products, just because they have adequate strength, should not be subject to hazards at too young an age. This growing maturity in pore structure is particularly relevant to cementitious mixes containing GGBS. These can take 3–6 months to achieve this improvement. The permeability in concrete containing GGBS can improve by an order of magnitude or more resulting in a concrete that is highly resistant to sea water and estuarine conditions.

9.7.3 Chemical

As with physical properties, chemical resistance generally relates to pore structure. In addition, the type of aggregate used comes into the picture. Limestone aggregates are soluble in most acids and it would seem logical to avoid their use in such conditions. However, the mechanism of attacking the surrounding matrix when insoluble aggregates are used can make matters worse. This leads to the promotion of the use of limestone as a sacrificial material promoting uniform rather than pitting corrosion.

Impurities in the aggregate can cause problems and, as recommended earlier, properties of all ingredients should be assessed before their use. Low alkali cements can be used when the aggregate has reactive material in it. Admixtures such as integral water repellents can suppress many other reactions and, in the case of quicklime-containing clinker, wetting and weathering before use can overcome the risk of later spalling.

9.7.4 Aesthetics

Probably enough has been said about this thorny problem and advice can be summarised in a few salient points for visual (architectural) precast products:

1   Construct at least two full-sized mock-ups incorporating all the reinforcement, insulation (if applicable), hardware, spacers, etc. that full production units will contain. These models should be made as if they were for full production and not treated as 'specials'. When agreed that these are the finishes that can be produced, one mock-up should remain in the works and kept in a suitable place in the works for comparison purposes and the other sent to the site. It would also be of benefit to include in both mock-ups the proposed on-site jointing and fixing specification so as to iron out any possible snags as early as possible.
2   If any finishing work is done on site, it should be restricted to washing and brushing with clean water. Acid should be avoided as well as any form of attrition such as stoning, grinding or grit-blasting.
3   Any making good or finishing work such as arris repair, filling blowholes, or cutting out pyrites iron oxide stains should be done in the works preferably in the first 48 hours.
4   Do not assume that the early appearance of a façade is what the building will necessarily look like when it ages.
5   Cleaning of buildings, when necessary or desirable, should be with a cold water supply from a rooftop sparge system with a back-up reservoir in case of mains failure.

10 Standards, testing and quality

10.1 General

It's a great pity that subjects involving documentation generally evince a mighty yawn with the common philosophy being that if the rulebook states that something has to be done, then the minimum of staff, finance and general interest will be deployed. How this philosophy manifests itself is that middle management staff are used full-time or part-time in quality-associated activities with technicians/operatives to support them. Not enough attention is paid to the head and the tail of this animal. At the head, lip service by managers is not enough and they need to get involved. This does not necessarily mean having meetings. A cursory interest will be seen as demotivating and managers should look deeper than the bottom line of a ledger to see the cost benefits of a total quality control system. At the tail end, staff and operatives will be encouraged to make a team effort to produce a product and service of which they can be proud. All those involved would also be encouraged to be proactive, put their ideas into the system, continually improve the processes and systems and, most importantly, form an appreciated part of the whole production team.

Much has been written by eminent authors on Standards, testing and quality, and here we discuss Standards, testing and quality in the precast industry only. The main reason for discussing the three subjects of the title in general terms is because each precast operation is specific to the local boundary conditions. Therefore, a general approach is given throughout.

10.2 Standards

Most of the control documents are in the form of ENs or EN ISOs with, at the time of writing, even some ISOs. In addition to these references, some BSs still obtain. Whatever the specific situation, it is essential to have hard copy of all specifications relevant to the business. As all these documents are copyright, they are not usually available in electronic format and may not be copied. However, (see Section 10.4) the specifics may be embodied into in-house 'procedure notes'.

As far as ENs and the like are concerned, the British Standard Institution has 'mirror' committees to each EC committee with the mandated responsibility to prepare a Standard. It is essential that the precast concrete industry through its Trade Associations and the Concrete Society represent their members' interests and attend both these BSI Committee meetings as well as those held in Europe. This is, apparently, an unpaid overhead (apart from fares and hotels for EC meetings) but needs to be accepted otherwise unwanted legislation might well appear that is due to the precast industry's lack of attention.

Where a Standard is specified and there is disagreement with part or all of its contents, this is irrelevant to the strict observation of the text and contents. The specified party keeps strictly to the letter of the Standard and neither looks for nor invents interpretation. If there is a need to amend a Standard, this is done through the British Standard committee and the European committee.

The application of Standards, testing and quality is not a cost and profit matter except in the most indirect terms. The situation is not all that dissimilar from exhibiting at a conference or a fair or attending a conference or meeting. It is an overhead that possibly boils down to a simple matter of making one's presence known and the accompanying publicity.

In addition to supporting activities at British Standard and European levels, Trade Federations and the Concrete Society offer opportunities to have one's say at national and regional events. The most important thing is to offer an opinion at the beginning of a development and not leave things until one is confronted by legislation that is faulty, irrelevant, illogical, or unacceptable.

The other documents that it may be necessary to hold are the relevant Regulations, NHBC Standards and Zurich Agreements, all of which are available electronically and may be downloaded from the web. In addition, there are 'supporting documents' which tend to be more relevant to construction rather than precast concrete manufacture but can still reflect on details in the units manufactured. The government sub-contracted the preparation of some of these documents to consulting organisations and it can take some time to trace these on the web. They are probably best found by typing in key words rather than a government document search.

The main problem encountered with all the above documents is that people are often too busy to give their views at drafting stages and are somewhat nonplussed when the publication stage is reached. Downtime in production for maintenance and cleaning has already been discussed and it is just as important in having one or more people in the works given enough downtime to keep an eye on document development and to be able to express views at the earliest opportunity.

At the time of publication of this book it was not mandatory to 'CE' mark goods for the market but it is predicted that a CE mandate will be in force in the UK shortly. What this means is that testing by an Approved Body to the requirements of a European Harmonised Standard (HEN) or a European Technical Approval (ETA) will be the two main routes to obtaining the mark. The CE mark is in common use in other Member States. The mark means that the activity or organisation placing the product or system on the market claims that it complies with the essential requirements of the specific Directive. It does not mean that the product or system is necessarily fit for purpose. It does not cover aesthetic requirements and is vague on the aspect of durability which, as a specific subject, is not covered in the Directive. It can also be argued that 'sustainability', a current theme word, is not addressed in any European documents.

What this means in practice is that National Regulations have to pick up the deficiencies in these head documents and cater for them by publishing National Annexes to them. To the author's knowledge, no National Annex has been published to an ETA but some of the essential British ENs have annexes. It is also of interest to comment that the number of interpretations in the Construction Products Directive led to the publication of Interpretative Documents, one for each of the six Directives. These Interpretative Documents dwarf in size the Directive on which they are based.

It's a great pity that so many mandatory documents seem to be plagued with anomalies and this is where precasters, serving on control national and European Committees, can have their say. At the receiving end of such a document, either within or referred to in a specificatation, it is suggested that before any activity is undertaken, a formal interpretation is established and documented both on the contract file and in a general contracts file. It is of little avail at a later stage arguing that so and so is what was intended.

A spin-off of all this is when the precaster's experience of good practice is at variance with what has been specified. This is a tricky situation and possibly all that can be done is to draw the client's attention to this formally (documenting all relevant notes) in that certain aspects can cause problems. It can be dangerous to recommend that what one did successfully on other occasions was another design but comment only that the alternative proved more amenable. All too often post-contractual interest is misinterpreted as precaster responsibility. This does not mean than satisfactory acceptance of products is a reason for not responding to later complaints. One's expertise can usually be used to highlight problems caused by others while, at the same time, maintaining a useful public relations exercise.

10.3 Testing

10.3.1 General

All too often testing is treated as a necessary appendage in the precast work whereas it has two main advantages, each of which excludes churning out quality control numbers:

1 a publicity centre showpiece of the factory operations;
2 a means of dealing with investigation and/or development.

Testing is generally used for ascertaining compliance and, in this respect, precast has an advantage over in situ concrete in that the product (realcrete) rather than a cube or cylinder (labcrete) is generally tested (see Chapter 4). Specifications for realcrete generally are not age-specified but, where this is the case, it tends to be at 28 days. This is of minimal interest to the precaster as it is the mechanical strength at the lifting, handling and stacking age which is typically in the range 6–16 hours.

When labcrete is tested, normally as cubes, it is essential to ensure that the production mirrors the actual process as closely as possible. Here, the manufacturer can develop in-house test relationships with non-destructive tests on the product, bearing in mind that such data is specific to that product and at the recorded age of both product and cube/cylinder test.

The reader is referred to the specific chapter in the materials book for a full discussion[68] and here the main considerations are summarised, not necessarily in order of their importance:

- Ensure that both documentation and traceability obtain.
- Ensure that the calibration status of equipment is correct.
- Mark 'Do not use' or similar if necessary.
- Report and act on non-compliances immediately.
- Give opinions on test procedures.
- Allow existing or potential visiting customers to have their say first without interruption other than asking for qualifications.
- Expect data to have a spread as concrete is not a constant.

10.3.2 Testing philosophy

The philosophy of testing concrete was discussed by the author in 1985.[69] The two main problems that exist in testing are either there singly or together and generally are rarely considered by the parties involved:

1 placing an almost religious trust in the test value;
2 assuming that the cube result related to all properties.

It wasn't until 1985 that a leading engineer published the realisation in the words: 'Our obsession with the cube has to go.'

The value of (1), the cube test, has already been discussed and, on (2) the ISAT test result received the comment on ISAT in 1965 that 'if the figure is beyond certain limits then the durability and the resistance to weathering are impaired'.[70] There are at least five bases, singly or jointly, addressing the 'Why test?' question:

1 The specification

This is a clause or clauses in the contract documents requiring maxima or minima figures for pass/fail. Tentative specification clauses such as 'in case of. . . . tests may need to be undertaken' should not be ignored as they could become bill items if invoked. Therefore, a price per test or similar approach should be formalised in the tender documents.

2 Quality control

Although this is discussed in fuller detail in the following section, it is worthy of mention here that quality control is the most justifiable reason for undertaking testing. Assuming that the test is performance-related, the data from such testing not only is a method of providing self-assurance in one's capabilities but also exhibits client-oriented publicity.

3 Curiosity

This aspect will only probably be the reason for testing where research and/or development are necessary. Remarks in previous chapters concerning the assessment of changes, e.g. a new source of aggregate, emphasise that the precaster is developing his state of knowledge by encompassing an extra variable in the data base.

4 Camouflage

Not a nice word to use but the author has come across examples in specifications where a test was scheduled whose value was completely irrelevant to performance with, at the same time, the omission of a relevant test. The example of this was a clause in a specification asking for the analysis of a red pigment used in the mix so as to inhibit the concrete from fading. The actual risk was masking by lime bloom which could have been assessed by a permeability test with a modified mix design.

5 Professionalism

This can be a somewhat unsavoury word in its interpretation and is where specifiers produce test clauses, often irrelevant to the product performance, to illustrate that they (the specifiers) are capable of calling up test requirements. This was illustrated earlier where the example was given of a freeze/thaw test for cast stone when there has never been a reported incidence of such damage.

Summing up this discussion, there are only two types of information required from testing precast concrete products (or in situ concrete as well) either singly or jointly: *STRENGTH* and *DURABILITY*.

Provided that all parties adopt an intelligent view and discuss matters at the pre-contract stage, then a sensible clause should appear in the specification. The author was rather pessimistic about such an action in 1985[71] and predicted that quality assurance had to be thrust upon the industry.

10.4 Quality

10.4.1 Names and definitions

Names such as 'Quality Assurance', 'Quality Control', 'Quality Management' abound in this particular subject and it is not surprising to find that the whole subject of quality tends to lose the reader's interest quite quickly. An attempt will be made in this section to hold this interest by placing the subject into as simple a context as possible, the key word being 'readability'. When one goes into detail of quality in industry and the many articles published on the subject as well as the conferences, seminars, etc. the author often finds it difficult to stifle a yawn. The basic precepts of applying a formal quality discipline to anything are, as the word states, 'basic'. The application is like the card game bridge. It is an easy game to learn but is made difficult by the people who play it.

The word 'quality' can be defined as: 'The totality of features and/or characteristics that define the ability of a product and/or service to meet a stated need.' The main document that one refers to in these matters is BS.EN.ISO.9001:2000[72] with its accompanying guide document BS.EN-.ISO.9004:2000. Parties claiming 'compliance' with 9001 do not tell the purchaser any more than that activity has an approved management system in place. There is no information on quality compliance levels or of poor, mediocre or good quality. The word 'quality' is a noun and not an adjective. Every food retailer sells 'quality food' and every auto-shop sells 'quality cars' irrespective of the level of quality.

There is also no reference in any of the published control documents to the need to improve continually the quality of the product or service. The main purpose is to improve continually the systems and processes that are used in the production. If the product is fit for purpose and acceptable, there is no need to improve it. To strive for improvement in the associated processes rather belies the saying 'If it isn't broken, leave it' when it should possibly be 'If it isn't broken, break it.'

The old ISO 9001–9004 series (which replaced the defunct BS 5750) were superseded by ISO 9001 and 9004 in 2000. Commercial vehicles can still be observed with defunct 'ISO 9002' imprinted thereon; this is certainly a questionable sign of that organisation's control when no such Standard currently exists. To attain a level where a product or system reaches the purchaser with an established supporting management structure involving many aspects including staff training, a suitable production and testing facility requires a total approach. Therefore, an overall better definition for all these activities is 'Total Quality Control'. Let us now discuss the items that make up this total.

10.4.2 Management

At the risk of being criticised, the whole subject of quality is simple but more often than not made complicated by the people involved, many of them being the professional practitioners. If one bears in mind that 'people' are more important than 'process', then things begin to fall into place quite neatly.

The quantity of downloaded documents produced, all too often is assumed to be a measure of quality. The documents are not necessarily related to total quality management. Management needs to accept several premises:

- There needs to be nominated person and a deputy with responsibilty for quality.
- That person must be assigned sufficient non-production time to oversee the systems efficiently.
- The management head document is a quality manual based upon the ISO 9001 headings.
- This head document must be flexible by referring to sub-documents that can be revised as necessary. The management personnel structure should feature as a sub-document.
- Relevant staff and operatives should be asked to comment on drafts of control/procedure documents and provide feedback on their use.
- Quality activities are a part of the production and are there to help not to hinder.
- Criticisms and comments should be encouraged. All staff and operatives are members of the quality team.

10.4.3 Control

To attain total quality control requires a staff/operatives team approach where every individual involved not only knows what they are supposed to do and achieve but also has an appreciation of the related tasks of their colleagues. Common sense should be applied in all matters; the 'just in time' approach used in some industries for goods-in is running production too close for comfort in precast concrete work. It is too easy to lose either part or all control because precast concrete production is a multi-variable staff/operatives/materials/plant juggling process and it is of benefit to have a 'Plan B' and, possibly, a 'Plan C' in all the systems, processes and goods.

11 Finishing, repairing and jointing

11.1 General

This chapter affords the opportunity for the author both to integrate the many papers he has published and to abstract from Concrete Society publications those aspects deemed relevant to precast concrete operations. Excluded from this chapter are any references to operations involving either poor quality concrete or upgrading good or mediocre quality concrete to withstand existing or additional hazards. This exclusion refers to activities involving paints, silicones and other surface/impregnation treatments and repairs to concrete damaged by reinforcement corrosion or other mechanisms.

11.2 Tools

There is a wide range of tools used in all the applications that are discussed later and what is used is not only the personal choice of the operative but, quite commonly, is a tool made specifically for the job. Therefore, only general guidance can be given and this is in the form of a list of recommendations.

For working on flat surfaces generally at least two tools are required, one to work in a filling material and one to rub back that surface to datum. Fine carborundum stones, hard sandstones and various types of hard natural stones as well as (for rubbing in) timber and acrylic. The proviso is that the tool used should neither be too soft so as to wear down too quickly nor too hard so as to wear down the surface too deeply. Commonly, visual concretes are under consideration and these often have a white or light-coloured cementitious base. Therefore, ordinary steel tools may well cause staining and this possible effect should be assessed in the selection of metal tools.

Profiled surfaces present some difficulty and the stone tool used may well need to be cut and ground to match the particular geometry. Most people tend to use their right hand rather than their left one and consideration needs to be given in tool selection when working from left to right on products that cannot be rotated to suit the work requirement.

Arrisses which require minor repair or 'sharpening' demand experience and dedication and can often be dealt with by running two stones at right angles to each other, either working in the filling material and then rubbing back to the datum levels when the filling material can be tooled. Sometimes, a stainless steel spatula or similarly-shaped tool with a right-angled cut in its nose suits some operatives. There are probably a number of other solutions to deal with arrisses and people can experiment under their own conditions to find out what suits them best.

Whatever tools are used for rubbing and filling work, the criteria to be generally met are:

1   Apply the material to fill blowholes and minor imperfections so that the filled surface is slightly proud of the datum target.
2   Rub back to restore datum and remove excess fill material when it has hardened.

Repairing and jointing work is not as demanding in tool selection but still requires expertise. For dealing with jointing of moulded faces, the surface laitance should be removed by tooling, grit-blasting or similar to expose a rough aggregate-visible surface.

The only occasion when the moulded surface may be left is when finishing gypsum plaster is to be applied directly to that surface using a dry condition polvinyl acetate (PVA) polymer-painted surface as a bond promoter.

Handling emulsion adhesives such as PVA, styrene butadiene rubber (SBR) and acrylics only needs standard tools such as brushes, trowels and the like which can be washed clean with water before the emulsion or polymer mortar has had a chance to dry and form a difficult-to-remove film or build-up on that tool. Paint brushes used in this work should be treated similarly and washed thoroughly. For all these emulsions a wash rota of about four per hour is a target that may be needed with a higher frequency in hot weather and a lower one when it is cold.

For repair work using thermoset resins such as polyester and epoxide, tools are as for the emulsions but either the supplier's bespoke cleaning liquid should be used or it may be found that placing the tools in water between uses may stop or inhibit the resin system hardening. Where very small brushes are used, it can be of benefit to use cheap disposable ones and not bother with cleaning or storing.

The other essential tool to have in the area of operation is an eye-washing proprietary station as well as a first aid kit. It is assumed that operatives will have been trained in the hazards of this work and supplied with protective clothing. They should also have been assessed for family and personal history of skin sensitivity. Fire-fighting equipment is also advisable for the thermoset resin polymers and, again, suppliers will advise and the local fire officers committee may well be able to visit the works and give the benefit of their expertise.

Disposal of polymer-containing tools and excess polymers and cleaning agents should be in the approved manner and should not contaminate domestic refuse or water courses. All polymers, chemicals and the like should also be stored in approved containers on which the suppliers can advise.

11.3 Materials and mixes

Materials can be:

(a) cementitious;
(b) fine aggregates;
(c) thermoplastics in the form of emulsions;
(d) thermoset plastics in the form of resins with accompanying catalysts and, sometimes, accelerators.

These can be used in any combination generally with the exception that cementitious-based mixes cannot be used as a rule with polyester or epoxide resins. There are some special epoxide resins into which Portland cement with water can be used and which harden by a redox process but such mixes are not included in the applications discussed later. A few words addressing each of these four groups of materials should be found helpful but, before venturing on any of the processes, in-works trials are necessary to ascertain what suits the application that one has in mind.

11.3.1 Cementitious

All cements and additives such as PFA and GGBS have been used successfully for many years and are usually mixed with fine aggregate and water to form a filling blend on very young surfaces or with PVA, SBR or PMMA emulsions for applications to concrete of a not young age, typically, over 2 days old.

Colour matching requires a somewhat alchemistic approach with one developing one's own recipes. There is only one possible general recommendation to bear in mind and that is that all non-white cement-based applications tend to lighten in colour as they become more mature whereas white and light-coloured cement-based ones tend to darken. In effect, whatever cementitious system is used, do not aim to achieve a colour match at the green stage.

11.3.2 Aggregates

Fine aggregate passing the 600 micron sieve is typically required for most applications emphasising that samples should be permanently retained for comparison purposes. Larger size gradings may be preferable for repairs and jointing work but, for filling application work, sizes coarser than the 600 sieve will often be found unsuitable for filling in very fine imperfections.

11.3.3 Thermoplastics emulsions

PVA, SBR and PMMA emulsions are often sold as about 30 per cent m/m solids and are diluted to a 10–15 per cent m/m emulsion with water to form the gauge for an aggregate/cement mix. These emulsions are film-forming and the film forms inside the thin mortar by air-drying of the water. PVA systems can be plasticised with chemicals such as dibutyl phthalate or maleic anhydride with the plasticising process being a molecular-internal one or, more commonly, an externally-plasticised one. What this means is that the common externally-plasticised resins can be 're-energised' after the film has hardened by contact with water or hydraulic-based systems. This can be a restriction on their use in exposed conditions but still allows the use of internally-plasticised PVAs which cannot be 're-energised' the same way. The latter system was used in repair work to the prestressed beams supporting the old Taylor Woodrow building in Southall, Middlesex, over the Grand Union Canal in 1960 and, when last seen by the author in the late 1990s, were unweathered.

The SBR and PMMA emulsion systems generally cannot be 're-energised' like the PVA systems and, once hardened, do not bond well if at all to following applications. Both the SBR and PMMA emulsions tend to lose water much quicker than the PVAs and the area being treated should be limited to a few square metres at a time, especially in warm and/or windy conditions. Mixing too large a quantity at a time can lead to unnecessary wastage.

Other thermoplastics polymer and copolymer systems are both available and in the development stages but changing or venturing into new chemicals can mean that one has to start a new track record of experience. Furthermore, those in current use have favourable track records of several decades and, in the author's opinion, do not need improvement. Any improvements required are in design for use as well as workmanship.

11.3.4 Thermoset polymers

These are commonly used as polyester or epoxide resins and set by chemical reaction between a resin and a catalyst where a cross-linking mechanism between virtually all the molecules in the resin is achieved. In effect, what was once billions of individual molecules can be traversed via atomic bonds from one end to the other. As an example, a 20m-long fibreglass boat is a single molecule.

The catalyst for polyester resins is in the form of a peroxide whereas, for an epoxide, it is an amine form. Personnel direct and family history of sensitivity to these chemicals should be ascertained before deploying staff/operatives in any of these activities. At all times, protective clothing should be worn, especially suitable gloves. Resin suppliers will advise in their health and safety literature.

Polyurethane resin systems can be used in coating applications and these are available as moisture-cured (relying upon atmospheric moisture to catalyse the system) as well as chemical catalyst cured. Their use in repair and jointing work is, as far as is known, not as common as for the polyester and epoxide resins.

Virtually all aggregates can be mixed with either polyester or epoxide resins. Some aggregates such as those from volcanic sources can be slightly acidic and can retard epoxide systems which prefer a slightly alkaline environment to harden efficiently. Polyester resins prefer a slightly acidic environment; this is why GRP moulds need to be suitably gel-coated and need treatment with a mould release agent to inhibit alkaline reaction with the cement in the mix.

11.4 Applications

The author appreciated the potential for these organic chemicals nearly 50 years ago and published what was probably the first UK's paper on the use of them in 1961.[73] This was discussed in fuller detail at a London Symposium in 1971[74] where fifteen applications were either described or proposed. Limitations on the fire-resistant properties of resins were studied and the results of that research were published in 1965.[75] Over the past two or three decades there have been countless publications, symposia, meetings and the like where the main concentration has been on repair and remedial work on concrete damaged by reinforcement corrosion. The applications that follow concentrate on the basic subject of making something good better, not on making something dubious acceptable for some time.

11.4.1 *Rubbing and filling*

For both visual concrete as well as the general trend of ordinary concrete, this is possibly one of the most common activities in the works. In finishing of visual concrete surfaces this part of the manufacture can take up a significant amount of the production time. The Empire Stone Company, probably the leading UK cast stone and general visual concrete manufacturer, used to spend about 50 per cent of production time on finishing work.

Depending upon the ambient curing conditions, this work should be undertaken at 16–36 hours old when the surface can be reworked. The surface is wetted first and a mix of about 1 to 2 parts of fine aggregate with 1 part of cement is rubbed into the surface using (see Section 11.2) an appropriate stone or tool. Care needs to be exercised when working near corners and arrises as these can be easily damaged at such a young age. Blowholes with a surface diameter above 3mm also need care as these can be undercut. What is visual as an approximate circular shape may well be over half a 'sphere' in size. If these are not fully filled, the stone or tool will drag the mortar, leaving a fine gap at the edge.

When fully rubbed in with the fill material, the finished datum should be 0.5–1.0mm proud of the design datum. This excess material should be allowed to harden for about 10–20 hours then rubbed back, working dry, with a suitable stone or tool. The blowholes will be visible as circles of mortar but, if the fill material is well selected (see Section 11.3), the colours will blend together after a short period of weathering.

Ex-mould wet-cast grey concrete vibrated product surface is not that amenable to this process as the surface is rich in fines and can be prone to crazing as well as showing other imperfections. It is suggested that clients be advised of this aesthetic drawback and be encouraged to specify an exposed-aggregate finish even though its depth would be almost pseudo-terrazzo in appearance.

11.4.2 Arris tightening

Either due to insufficient compaction and/or to disturbance during demoulding arrises can become damaged or deficient in material. Following the same recommendations of both mix and timings as well as pre-wetting, apply the repair material using two tools combined or two stones working at right angles to each other and work the material along each affected edge. However, do not use excess material so that the wet-finished material is the design datum level. An alternative to using two stones is the spatula with a right-angled cut in its nose.

When the material is hard, gentle abrasion with a dry stone, sandpaper or fine carborundum will reinstate the required appearance.

In some instances both processes described in Sections 11.4.1 and 11.4.2 need to be undertaken contemporaneously. In such cases, since the surface datum has to be raised so does that at the arrisess. Therefore, the dry rubbing process in the hardened stage will need to be more rigorous at the arrisses, often using stones or tools with a rougher surface than would be used for arris tightening alone.

11.4.3 Blowhole and honeycomb filling

This is a process that is more often than not applied to ex-mould ordinary concrete where the remaining surface is acceptable. Using a square-ended piece of hardwood, push in a fine mortar mix typically 2 parts of fine 'sand' (natural or crushed rock or stone) to one part of cement with water added to form a thick cream. The surface should be slightly dampened first but there should be no free-sitting water inside the blowholes or the honeycombed areas. The mix used should generally be darker in appearance than the surrounding matrix as it will lighten to give a fair match on weathering.

After the filling process, use a damp rag as soon as possible to remove excess mortar/grout from the ex-mould surrounding area. There should be no need for a dry-rubbing process the following day.

Honeycombed areas should be first rigorously tested with a cold chisel or similar tool to ensure that the area designed to be repaired is sound enough to support the repair material. If not, the area in question should be treated as described in Section 11.4.4.

11.4.4 Spalled area repairs

This method applies to areas spalled generally due to accidental damage as well as to honeycombed areas. All these areas have to be excavated to a sound base. Preparation is of prime importance in this application and it is necessary to do the following:

1   avoid feather-edging by cutting into the periphery at least 10mm deep to form a turret or reverse mitre cut;
2   ensure that there is a slight mechanical key at the base; too deep a key might well inhibit good bonding.

Having prepared the area to receive the repair material, the following procedure has been found suitable although other polymer systems may be suitable. That recommended here has a good track record covering over four decades to date with continuing good performance:

1   Wet the area(s) and remove free-standing water.
2   Using a matching cement, mix up a grout with an SBR emulsion of about 15 per cent solids and apply to the receiving surfaces with a brush, working it well into all undulations.
3   While this grout is wet/tacky but definitely not hardened, mix up a moist mix design mortar of about 2 parts of fine aggregate to 1 part of cement using the same emulsion as for the grout and rigorously ram the mix into the area and then trowel smooth.
4   Rub back to datum when the repair material has hardened.

NOTE: polymer repairs generally lighten as they dry and harden and do not darken under wet conditions. Matching repairs is very difficult and a dry surface match will not be a wet surface match. An advanced remedial technique to achieve a better match if the spalled area is deep enough to receive a two-layer repair mortar is to finish the SBR mortar about 10mm below datum, comb the fresh SBR mortar surface and trowel on a moist mix design mortar not containing polymer. This technique requires considerable practice to achieve acceptable results.

11.4.5 Hydration staining

This procedure is almost identical to that described in Section 11.4.4 but it has been stated earlier that this staining can penetrate 30mm or more in depth, therefore there is the added procedure of dealing with exposure of reinforcement. However, the 'advantage' of such exposure is that this is not a chloride corrosion situation and therefore only the exposed steel needs attention so the following method may be used:

1   Wet the area(s) and remove free-standing water.
2   As step (2) in Section 11.4.4 but ensure that all exposed steel surfaces receive the grout; this may involve the use of special-shaped brushes.
3   and 4 as (3) and (4) in Section 11.4.4.

The same note also applies with an added warning. Hydration staining repair can involve large areas and large amounts of repair material and the warning about not mixing more than can be used in the pre-drying period is repeated.

11.4.6 Pyrites staining

The iron sulfide (pyrites) particles often found in sedimentary flint, limestone and intermediary gravels have always caused a severe aesthetic problem. The effect of the atmosphere on the particle is to produce iron oxide rust which stains the area of the particle and penetrates the surface. The product affected is, more often than not, a visual cladding unit which is probably stored vertically. Therefore, the stain runs down the face with the offending particle(s) at the head of the stain.

What this means is that the quicker one observes and actions the defect the better. Therefore, regular stack inspections are advisable with enough room between units to be able to pick out these stains. This, in turn, means that most of the remedial work needs to be done in the works. The remaining problem arises from particles resting a few millimetres below the surface which can necessitate remedial work on site as staining is likely to manifest itself after delivery.

From the contractual viewpoint if a client specifies a particular aggregate, he/she should be formally advised of a risk. If, on the other hand, the aggregate is the precaster's selection, then he/she should attempt to purchase material from zero or low risk sources.

The procedure is as Section 11.4.4 with the added recommendation of going for a two-layer repair if the area is large enough, say, over 500 mm^2.

With exposed aggregate finishes hand placement of stones may be necessary, again, an application needing a lot of practice to achieve a good match. The stained area can either be cut out, preferably, or treated with a 'stain remover' as described for tackling rust stains.[76] This chemical method involves the use of a sodium citrate/glycerine poultice in whiting or kieselguhr carrier with possible repeat applications. With the best will in the world it is virtually impossible to achieve solution of the iron oxide and a 'ghost stain' is left. After all, the iron oxide stain is the same chemical as that used for pigmenting concrete and it needs little imagination to consider how difficult or impossible it would be to deal with such staining.

11.4.7 Accidentally-cracked reinforced concrete

This method was described over 40 years ago[77] and applies to reinforced concrete units cracked due to an accident such as dropping, handling too young or stack overloading or similar. The procedure is as follows:

1   Store unit in warm, dry conditions for at least a week.
2   If necessary, constrain unit until the crack has a minimum surface width of 0.4mm round its periphery.
3   Place on supports and jack as necessary to achieve orthogonality in all directions ensuring, also, that no support obscures the crack at its soffit.
4   Using a slightly greased 'Plasticine' (some types of adhesive tape might also be suitable), seal off the crack at the soffit and the vertical sides.
5   To the manufacturer's instructions mix up a low viscosity polyester resin and apply to the centre of the top surface of the crack until refusal.
6   When hardened, remove the sealing material.
7   When appearance is a consideration, a fine mortar repair as described in Section 11.4.1 may be used.
8   For added fire resistance, not that resin in a fine crack has been shown to be a risk, the repair crack line can be cut back to form a trench about 10mm wide and deep and made good with a matching (when hardened and dry) non-polymer limestone fines/cement mortar.

Polyester resins normally will achieve strengths over that of the concrete within a few minutes to a few hours depending upon the type of polyester resin and the quantities and types of hardener and accelerator. Repaired units may be stacked at an early stage. The repaired cracked unit just described can be flexurally loaded for proof of integrity of the repair but the unit would almost certainly become scrap if loaded to failure. The new crack formed at failure will be in the region of maximum ultimate strain and incipient invisible cracking in that area will be present. A secondary repair is thus ill-advised.

11.4.8 Joining units together

This procedure mainly applies to the joining of reinforced concrete units together end to end. The application arose from an initial academic study to prove that resin adhesion of concrete was stronger and faster than any concrete in general production. The first demonstration was to stick together the broken faces of a three-point flexurally-tested 100 × 100 × 500mm beam and retest it when the resin had hardened. In every case the retest failure was in a new position. This was not surprising as a three-point test causes an ultimate failure zone over a distance of more than 130mm, therefore the test only proved that the polyester resin stuck the broken faces together but not how strongly. The logical test for the resin strength was to break the beam in flexure, wire brush or sandblast the end faces and turn the two halves sticking the butt ends together then retesting when the resin had hardened. As described in the publication,[78] the beam broke remote from the adhered ends.

Industrial application of this procedure soon followed at Gatwick Airport where a half ridge hangar roof was being constructed in order to produce a 'half' building enabling light aircraft to taxi to an undercover passenger area. The original planned construction was to post-tension the rafter to a seating in the column using a sand cement mortar in the joint. The rafter was supported on falsework until the mortar had hardened sufficiently to take the prestress force. This period, as determined by the mortar cube strength, varied from 4 to 7 days. The idea was put to the designer to use a polyester resin mortar in order to speed things up. The relative cost of this mortar compared to that of the sand cement elicited a pessimistic response which was completely reversed when the engineer in question realised that the prestressing process could be accelerated from one every four days to four a day. The rental cost savings in the falsework as well as the accelerated speed of construction more than offwrote the increased material cost.

The step-by-step procedure was as follows:

1   Wire brush, sandblast or pretreat jointing mould faces with retarder and expose aggregate at the demoulding stage.
2   Support rafter on falsework and insert suitable thin tubing into the duct hole ends to prevent resin mortar entering duct.
3   Dry mating faces if necessary.
4   Mix up polyester resin with catalyst and (if necessary) accelerator adding fine aggregate to form a thick cream consistency and spread onto both faces.
5   Bring the two faces together exuding excess resin mortar, removing this before it sets. The prestress cable can be used to apply this slight compression but no undue force is necessary.
6   Apply design prestress force an hour or two later when the mortar has hardened.

Note: For added fire resistance the joint can be chased out or the units step-jointed to allow the later application of a conventional mortar.

There are many uses for jointing precast units together and one used successfully by a cladding manufacturer was a case of a white exposed aggregate concrete panel with a nominally black panel set at a right angle at its end to form an L-shaped in plan unit. However, the author suggested another end-joint application to a precast pile manufacturer where several precast piles each about 20m long were needed for a London contract. It was alleged that no manufacturer wanted the job because of the problem of delivery through some very narrow streets in the dockland area. Therefore, in order to remain on the tendering list all tenders were submitted at very high levels and the lowest of these high price tenders got the job. The piles were delivered to the site but the manufacturer had to pay considerable damages for impact damage to property and much street furniture.

The suggestion made at that time but not taken up was to make three piles for series driving for the same length as the original one designed. The reinforcement in the 'receiving' pile would have metal tubes welded to the ends finishing flush with the end surface. The 'offered' pile would have protruding reinforcement to match the tubing on a 'plug and socket' design. After driving in the first pile and appropriate cleaning and drying of the mating faces the metal tubes would be filled to a suitable depth with a neat polyester resin and the second pile offered up with excess resin spilling from the joint. When hardened, the second pile would be driven in, pushing the first pile ahead. The process could then be repeated for the third series pile. Obviously there would need to be engineering consideration as to the profiles of the tubing and the protruding reinforcement.

11.4.9 Jointing masonry units having extreme suctions

Products having integral water repellents in the mix are difficult to mortar as the surface is water repellent and mortar beds do not adhere well and fine line cracking can result, coupled with possible poor bonding. At the other end of the spectrum concrete blocks such as autoclaved aerated ones have high suctions to mortar and remove water from the mortar mix before hardening can start. On site some of this effect can be alleviated by wetting the blocks but this can cause excess moisture movement shrinkage as the blockwork dries out.

The remedy to both low and high suction cases is the same mortar but for different reasons. The mortar suggested is a cellulose polymer system. For low suction surfaces the cellulose has a significant effect on reducing the surface tension which, in turn, improves the wettability. This can be easily observed in the laboratory or on site by the angle of repose of a dab of mortar on the test surface. This angle will be quite steep for the control mortar but shallow for the cellulose one.

The behaviour of a cellulose mortar on a high suction surface has little to do with surface tension but with movement of water within the long chain polymer system of the mortar. This means that the masonry has little suction effect on the mortar. In addition, both for low and high suction surfaces the face of the mortar joint can be reworked up to an hour or two even in very hot dry weather.

Care needs to be taken on site as celluloses are retarders and masonry should not be built up too high too quickly otherwise instability could result.

11.4.10 Surfacing

There are several applications of both thermoplastics and thermoset polymers worthy of mention but the ones listed are not claimed to be exhaustive. The common denominator in all these applications is that the polymer is never used neat; there is always aggregate with or without cement. The reason for this is that degradation due to the effect of ultraviolet light can occur. Even though the aim is to produce the longest-lasting application possible, maintenance work is easier than with the aforementioned applications as the areas are accessible.

The surface preparation procedure is generally the same for any of the applications; it is the application that varies. For concrete any laitance needs to be removed by suitable means. This could be by the use of sand or gritblasting or by the use, during manufacture, of a fine-acting surface retarder on the mould. Acid etching is not recommended.

Surfacing natural stone

This method has been used on a natural Portland Stone faced building where the façade had a mixture of Whitbed with a few Roach stones. A mix of 2 parts of Portland Stone fines with one part of white cement containing about 5 per cent m/m vinyl versatate polymer powder was sent to site as a prepack. To this water was added to form a workable cream which was rubbed into the surface with a fine carborundum stone. The following day the surface was dry-rubbed back to the original datum. The appearance was (and still is) so good that the architect at that time thought that the contractor had removed an unsightly Roach stone and replaced it with a Whitbed variety.

The polymer used was a versatate which is available in powder form and is suitable for prepacks. The polymer concentration was deliberately lower than for the type applications described in Section 11.4.1 as the aim was to promote similar weathering properties to the natural surfaces. The repair contractor used the prepack to precast on site dentils with protruding dowels as many of the existing ones had become too badly weathered to repair. A couple of simple wooden moulds were made and the mix was of a low water content and rammed into the moulds using hand tools.

Non-slip surfaces

This application is suitable for steps, cold storage room floors and products where health and safety is an issue. To the prepared surface apply a polyester or epoxide resin and, while the resin is unset, broadcast a nominally single-sized carborundum or other suitable hard-wearing aggregate so that there is excess material present. When the resin has hardened, brush off and recycle the excess.

Reflective surfaces

This is the same procedure as described in Section 11.4.10 except that the products and the material used are different. The products could typically be kerbs, safety barriers, lighting columns, bollards and similar. The aggregate used for broadcasting onto the resin-coated surface could be a white calcined flint, white alumina particles or similar. The mica particles in Cornish 'sand' granite fines have very good reflective properties and could be assessed either as an additional material to one of the above or even tried alone. The addition of rutile titanium oxide white pigment to white calcined flint is not recommended as the surface tends to have a high and visually uncomfortable solar reflection.

Notes and references

Where a book is referenced the reader is referred to the index of that book for the relevant text. All BS.EN and BS referenced documents are published by the British Standards Institution in London. All 'CSTR', 'CS' and similar initials refer to publications by the Concrete Society, Camberley, Surrey, UK.

1 Levitt, M. (1982) *Precast Concrete Materials: Manufacture, Properties and Usage*. Andover: Applied Science Publishers.
2 BS.EN.206–1:2004. Concrete – Part 1: Specification, production and conformity.
3 BS.EN.1990:2006. Eurocode – Basis of structural design (Part 1.1 and its relevant parts).
4 BS.8500:2000. Complementary British Standard to BS.EN.206–1. Part 1: Method of specifying and guidance for the specifier. Part 2: Specification for the constituent materials of concrete.
5 BS.EN.197–1:2004. Cements – Part 1: Composition, specification and conformity criteria for common cements (also its relevant parts). BS.EN.14216–4: Criteria for very low heat cements.
6 Levitt, M. (1997) *Concrete Materials: Problems and Solutions*. Andover: E & FN Spon.
7 Ibid.
8 Ibid.
9 Smith, M.R. and Collis, L. (1993) *Aggregates: Sand, Gravel and Crushed Rock Aggregates for Construction Purposes*. London: The Geological Society.
10 Clarke, J.L. (1993) *Structural lightweight aggregate concrete*. Andover: Chapman & Hall.
11 The Stationery Office (1995) *Use of Waste and Recycled Materials as Aggregates: Standards and Specifications*. London: HMSO.
12 CSTR18:2001. A guide to the selection of admixtures for concrete.
13 Ramachandran, V.S. (1995) *Concrete Admixtures Handbook: Properties, Science and Technology*. NJ: Noyes Publications.
14 Levitt (1982), op. cit.
15 CSTR40:2000. The use of GGBS and PFA in concrete.
16 Swamy, R.N. (1991) *Blended Cements in Construction*. Andover: Elsevier Science Publishers.
17 Ramachandran (1995), op. cit.
18 Owens, P.L. and Buttler, F.G. (1980) 'The reaction of fly ash and Portland cement with relation to the strength of concrete as a function of time and temperature',

paper presented at 7th International Congress on the chemistry of cement. Vol. III *Communications*. Paris.

19 Ibid.

20 Levitt, M. (2001) 'Silane surfeit on the Second Severn Crossing and other concretes', *The Structural Engineer*, 79(5): 14–16.

21 CSTR No.41. (1993) Microsilica in concrete.

22 Ramachandran (1995), op. cit.

23 Levitt, M. (1985) 'Pigments for concrete and mortar', *Current Practice Sheet*, 99 *Concrete*, March, 21–22.

24 Levitt (1997), op. cit.

25 UK CARES Sevenoaks, Kent. Website for latest update, Technical Information.

26 BS.EN.10080:20054: Steel for the reinforcement of concrete-weldable reinforcing steel-general.

27 BS.6744:2001. Stainless steel bars for the reinforcement of and use in concrete – requirements and test methods.

28 BS.EN.10088–1:2005. Stainless steels – Part 1: List of stainless steels.

29 Levitt (1997), op. cit.

30 Levitt (1982), op. cit.

31 BS.EN.14845–1:2003 (Draft). Test methods for fibres in concrete. Part 1: Reference concretes.

32 BS.EN.14845–2:2003 (Draft). Test methods for fibres in concrete. Part 2: Effect on concrete.

33 BS.EN.14889–1:2004 (Draft). Fibres for concrete. Part 1: Steel fibres: definitions, specifications and conformity.

34 BS.EN.14889–2:2004 (Draft). Fibres for concrete. Part 2: Polymer fibres: definitions, specifications and conformity.

35 BS.EN.14649:2005. Precast concrete products: test method for strength retention of glass fibres in cement and concrete.

36 BS.EN.ISO.15630–3:2002. Steel for the reinforcement and prestressing of concrete-test methods. Part 3: Prestressing steel.

37 Ibid.

38 Levitt, M. and Herbert, M.R.M. (1970) 'Performance of spacers in reinforced concrete', *Civil Engineering*, August.

39 CS101. (1989) *Spacers for Reinforced Concrete*. Camberley: The Concrete Society.

40 BS.7973–1(2001). Spacers and chairs for steel reinforcement and their specification. Part 1: Product performance requirements.

41 BS.7973–2(2001). Spacers and chairs for steel reinforcement and their specification. Part 2: Fixing and application of spacers and chairs and tying of reinforcement.

42 Levitt and Herbert (1970), op. cit.

43 BS.EN.15113–1:2005 (Draft). Vertical formwork. Part 1: Performance requirements, general design and assessment.

44 Levitt (1997), op. cit.

45 Richardson, J.G. (1991) *Quality in Precast Concrete*. Harlow: Longman Scientific & Technical Publishers.

46 BS.EN.10088–1:2005. Stainless Steels Part 1: List of stainless steels.

47 See Levitt (1997), op. cit.

48 The Prestressed Concrete Institute (1994) *Architectural Precast Concrete*. Chicago: The Prestressed Concrete Institute.
49 CS156:2006. The role of water in concrete and its influence on properties.
50 BS.EN.197–1:2004 Cements Part 1: Composition, specification and conformity criteria for common cements.
51 BS.EN.14216–4: Criteria for very low heat cements.
52 CS126:1998. In situ concrete strength-cube/core relationship (Coupled with Project Report. 3 Report of a Working Party of The Concrete Society, 'In situ concrete strength: An investigation into the relationship between core strength and standard cube strength' (2000).
53 Levitt (1997), op. cit.
54 CSTR31:1992. Permeability testing of site concrete.
55 Ibid.
56 Levitt, M. (1985) 'The ISAT: A non-destructive test for the durability of concrete', *British Journal of Non-Destructive Tests*, 13(4).
57 BS1217:1997. Cast stone.
58 CS126:1998. 'In situ concrete strength', op. cit.
59 BS.EN.771–5:2005. Specification for masonry units. Part 5: Manufactured stone masonry units.
60 BS1217:1997. Cast stone.
61 BS.EN.771–5:2005 Specification for masonry units.
62 Levitt (1997), op. cit.
63 Ibid.
64 The Prestressed Concrete Institute (1994), op. cit.
65 Levitt (1997), op. cit.
66 BS1217:1997. Cast Stone.
67 Levitt (1997), op. cit.
68 Ibid.
69 Levitt, M. (1985) 'The philosophy of testing concrete', *Concrete*, December 1985: 4–5.
70 Levitt (1985), op. cit.
71 Ibid.
72 BS.EN.9001–1:2000. Quality Management Systems Requirements.
73 Levitt, M. (1961) 'The use of adhesives in the bonding and repair of precast products', *Civil Engineering*, March and April: 1–4.
74 Levitt, M. (1971) 'Plastics with concrete', Paper presented at Plastics Institute Symposium Plastics in Construction April 1971, Polytechnic of the South Bank, London.
75 Levitt, M. (1965) 'The fire-resistance of resin-jointed concrete', paper presented at Conference Plastics in Building Structures, 14–16 June 1965, The Plastics Institute, London.
76 Levitt (1982), op. cit.
77 Levitt (1961), op. cit.
78 Ibid.

Web addresses

www.bsi-global.com.uk – British Standards Institution
www.concrete.org.uk – the Concrete Society.

Index